Hands-On Data Analysis with Scala

Perform data collection, processing, manipulation, and
visualization with Scala

Rajesh Gupta

BIRMINGHAM - MUMBAI

Hands-On Data Analysis with Scala

Copyright © 2019 Packt Publishing

Acquisition Editor: Yogesh Deokar
Content Development Editor: Unnati Guha
Technical Editor: Sayli Nikalje
Copy Editor: Safis Editing
Project Coordinator: Manthan Patel
Proofreader: Safis Editing
Indexer: Pratik Shirodkar
Graphics: Jisha Chirayil
Production Coordinator: Jisha Chirayil

First published: April 2019

Production reference: 1020519

Published by Packt Publishing Ltd.
Livery Place
35 Livery Street
Birmingham
B3 2PB, UK.

ISBN 978-1-78934-611-4

www.packtpub.com

I dedicate this book to Shalini, Krish, and my beloved parents.

`mapt.io`

Mapt is an online digital library that gives you full access to over 5,000 books and videos, as well as industry leading tools to help you plan your personal development and advance your career. For more information, please visit our website.

Why subscribe?

- Spend less time learning and more time coding with practical eBooks and Videos from over 4,000 industry professionals

- Improve your learning with Skill Plans built especially for you

- Get a free eBook or video every month

- Mapt is fully searchable

- Copy and paste, print, and bookmark content

Packt.com

Did you know that Packt offers eBook versions of every book published, with PDF and ePub files available? You can upgrade to the eBook version at `www.packt.com` and as a print book customer, you are entitled to a discount on the eBook copy. Get in touch with us at `customercare@packtpub.com` for more details.

At `www.packt.com`, you can also read a collection of free technical articles, sign up for a range of free newsletters, and receive exclusive discounts and offers on Packt books and eBooks.

Contributors

About the author

Rajesh Gupta, is a hands-on big data tech lead and enterprise architect with extensive experience in the full life cycle of enterprise grade software development. He has successfully architected, developed, and deployed highly scalable data solutions using Spark, Scala, and the Hadoop technology stack for several US-based enterprises. A passionate, hands-on technologist, Rajesh has master's degrees in mathematics and computer science from BITS, Pilani (India).

About the reviewer

Manoj Kumar, is an experienced consultant with more than 16 years of versatile experience across a variety of environments, including exposure to implementing process improvement and operation optimization in typical manufacturing environments and production industries using advanced predictive and prescriptive analytics, such as machine learning, deep learning, symbolic dynamics, neural dynamics, circuit mechanisms, and the Markov Decision Process.

His domain experiences are in transportation and supply chain management, processes and manufacturing, mining and energy, retail, CPG, healthcare, marketing, and F&A.

Packt is searching for authors like you

If you're interested in becoming an author for Packt, please visit `authors.packtpub.com` and apply today. We have worked with thousands of developers and tech professionals, just like you, to help them share their insight with the global tech community. You can make a general application, apply for a specific hot topic that we are recruiting an author for, or submit your own idea.

Table of Contents

Preface

Efficient business decisions with an accurate understanding of business data help to deliver better performance across products and services. This book will help you to leverage the popular Scala libraries and tools to perform core data analysis tasks with ease.

The book begins with a quick overview of the building blocks of a standard data analysis process. You will learn how to perform basic tasks such as the extraction, staging, validation, cleaning, and shaping of datasets. You will later deep dive into the data exploration and visualization areas of the data analysis life cycle. You will make use of popular Scala libraries such as Saddle, Breeze, and Vegas to process your datasets. You will learn statistical methods for deriving meaningful insights from data. You will also learn how to create applications for Apache Spark 2.x on complex data analysis, in real time. You will discover traditional **machine learning** (**ML**) techniques for doing data analysis.

By the end of this book, you will be capable of handling large sets of structured and unstructured data, performing exploratory analysis, and building efficient Scala applications to discover and deliver insights.

Who this book is for

If you are a data scientist or a data analyst who wants to learn how to perform data analysis using Scala, this book is for you. All you need is knowledge of the basic fundamentals of Scala programming.

What this book covers

Chapter 1, *Scala Overview*, gives you a quick run through Scala and its features. It will prepare you for upcoming chapters.

Chapter 2, *Data Analysis Life Cycle*, turns the focus exclusively to data analysis and its typical life cycle. It provides an overview of the steps involved in the data analysis life cycle.

Chapter 3, *Data Ingestion*, deep-dives into the data ingestion aspects of the data life cycle. It covers extraction, staging, validation, cleaning, and shaping data tasks. It highlights how to deal with the variety aspect of data, that is, how to handle data from different sources in different formats.

Chapter 4, *Data Exploration and Visualization,* deep-dives into the data exploration and visualization parts of the life cycle. It familiarizes the reader with techniques for discovering inherent properties associated with data using statistical as well as visual methods.

Chapter 5, *Applying Statistics and Hypothesis Testing,* provides an overview of the statistical methods used in data analysis and covers techniques for deriving meaningful insights from data.

Chapter 6, *Intro to Spark for Distributed Data Analysis,* covers the transition to doing data analysis on distributed systems and doing it at scale. It provides a good introduction to Spark, a Scala-based distributed framework for data processing. It will guide you through Spark setup on your computer and introduce key features using practical examples.

Chapter 7, *Traditional Machine Learning for Data Analysis,* covers topics such as decision trees, random forests, lasso regression, and k-means cluster analysis. It also covers the role of NLP in effectively analyzing certain types of data.

Chapter 8, *Near Real-Time Data Analysis Using Streaming,* introduces the concept of stream-oriented processing and compares it to traditional batch-oriented processing. It also illustrates how streaming can be used to perform near real-time data analysis. This chapter deep-dives into Spark Streaming and will guide you on implementing clustering and a classifier leveraging Spark Streaming APIs.

Chapter 9, *Working with Data at Scale,* is dedicated to processing data at scale. It looks at data analysis from multiple dimensions, such as cost, reliability, and performance. It provides guidance on some of the best reliability and performance practices. It provides a complete picture of how a practical real-world data analysis life cycle works and will help you to put this into practice in a production environment.

To get the most out of this book

- You should be familiar with the fundamentals of the Scala programming language
- You should have a passion for analyzing data and extracting insight from of it
- You should have basic familiarity with statistical methods and machine learning algorithms

Download the example code files

You can download the example code files for this book from your account at
`www.packt.com`. If you purchased this book elsewhere, you can visit
`www.packt.com/support` and register to have the files emailed directly to you.

You can download the code files by following these steps:

1. Log in or register at `www.packt.com`.
2. Select the **SUPPORT** tab.
3. Click on **Code Downloads & Errata**.
4. Enter the name of the book in the **Search** box and follow the onscreen instructions.

Once the file is downloaded, please make sure that you unzip or extract the folder using the latest version of:

- WinRAR/7-Zip for Windows
- Zipeg/iZip/UnRarX for Mac
- 7-Zip/PeaZip for Linux

The code bundle for the book is also hosted on GitHub
at `https://github.com/PacktPublishing/Hands-On-Data-Analysis-with-Scala`. In case there's an update to the code, it will be updated on the existing GitHub repository.

We also have other code bundles from our rich catalog of books and videos available
at `https://github.com/PacktPublishing/`. Check them out!

Download the color images

We also provide a PDF file that has color images of the screenshots/diagrams used in this book. You can download it here: `https://www.packtpub.com/sites/default/files/downloads/9781789346114_ColorImages.pdf`.

Conventions used

There are a number of text conventions used throughout this book.

`CodeInText`: Indicates code words in text, database table names, folder names, filenames, file extensions, pathnames, dummy URLs, user input, and Twitter handles. Here is an example: "Create a package called `handson.example` by expanding
to `src/main/scala` and right-clicking on this folder."

A block of code is set as follows:

```
scala> def factorial(n: Int): Long = if (n <= 1) 1 else n * factorial(n-1)
factorial: (n: Int)Int

scala> factorial(5)
res0: Long = 120
```

Any command-line input or output is written as follows:

```
$ brew install sbt@1
```

Bold: Indicates a new term, an important word, or words that you see onscreen. For example, words in menus or dialog boxes appear in the text like this. Here is an example: "Click on **Create New Project**, and then click on **Scala** and select the **sbt** console."

 Warnings or important notes appear like this.

 Tips and tricks appear like this.

Get in touch

Feedback from our readers is always welcome.

General feedback: If you have questions about any aspect of this book, mention the book title in the subject of your message and email us at customercare@packtpub.com.

Errata: Although we have taken every care to ensure the accuracy of our content, mistakes do happen. If you have found a mistake in this book, we would be grateful if you would report this to us. Please visit www.packt.com/submit-errata, selecting your book, clicking on the Errata Submission Form link, and entering the details.

Piracy: If you come across any illegal copies of our works in any form on the Internet, we would be grateful if you would provide us with the location address or website name. Please contact us at copyright@packt.com with a link to the material.

If you are interested in becoming an author: If there is a topic that you have expertise in and you are interested in either writing or contributing to a book, please visit `authors.packtpub.com`.

Reviews

Please leave a review. Once you have read and used this book, why not leave a review on the site that you purchased it from? Potential readers can then see and use your unbiased opinion to make purchase decisions, we at Packt can understand what you think about our products, and our authors can see your feedback on their book. Thank you!

For more information about Packt, please visit `packt.com`.

Section 1: Scala and Data Analysis Life Cycle

In this section, you will gain an insight into what data is, how it is prepared and processed, and how it is analyzed and stored. This section will also get you introduced to the Scala framework and how we use Scala with data. You will learn some basic commands in Scala. You will also understand what a data pipeline is and the tasks involved in this pipeline.

This section will contain the following chapters:

- Chapter 1, *Scala Overview*
- Chapter 2, *Data Analysis Life Cycle*
- Chapter 3, *Data Ingestion*
- Chapter 4, *Data Exploration and Visualization*
- Chapter 5, *Applying Statistics and Hypothesis Testing*

1
Scala Overview

Scala is a popular general-purpose, high-level programming language that typically runs on the **Java Virtual Machine (JVM)**. JVM is a time-tested platform that has proven itself in terms of stability and performance. A large number of very powerful libraries and frameworks have been built using Java. For instance, in the context of data analysis, there are many Java libraries available to handle different data formats, such as XML, JSON, Avro, and so on. Scala's interoperability with such well-tested libraries helps increase a Scala programmer's productivity greatly.

When it comes to data analysis and processing, it is often the case that there is an abundance of data transformation tasks that need to be performed. Some examples of such tasks are mapping from one representation to another, filtering irrelevant data, and joining one set of data with another set. Trying to solve such problems using the object-oriented paradigm often means that we have to write a significant amount of boilerplate code even to perform a fairly simple task. Oftentimes, solving data problems requires thinking in terms of input and transformations that are to be applied to this input. Scala's functional programming model provides a set of features that facilitate writing code that is concise and expressive. Spark is a popular distributed data analytics engine that has almost entirely been written in Scala. In fact, there is a strong resemblance between the Scala collection API and the Spark API.

Most of the Java libraries can be used with relative ease from Scala code. One can easily mix object-oriented and functional styles of programming in the same Scala code base. This ability provides a very simple pathway to a great deal of productivity. Some of the major benefits of using Scala are as follows:

- Most Java libraries and frameworks can be reused from Scala. Scala code is compiled into Java byte code and runs on JVM. This makes it seamless to use Java code that has already been written from a Scala program. In fact, it is not uncommon to have a mix of both Java and Scala codes within a single project.
- Scala's functional constructs can be used to write code that is simple, concise, and expressive.
- We can still use object-oriented features where they are a better fit.

There are many useful data libraries and frameworks that are built using Scala. These are summarized later in this chapter. Apache Spark needs a special mention. Apache Spark has become a de facto standard for performing distributed data analysis at scale. Since Spark is almost entirely written in Scala, its integration with Scala is the most complete, even though it has support for Java, Python, and R as well. Spark's API has been heavily influenced by Scala's collection API. It also leverages Scala's case class features in its dataset API and significantly helps in reducing the writing of boilerplate code that is otherwise necessary for Java.

The following topics will be covered in this chapter:

- Installing and getting started with Scala
- Object-oriented and functional programming overview
- Scala case classes and the collection API
- Overview of Scala libraries for data analysis

Getting started with Scala

At the time of writing, Scala Version 2.12.8 (`https://www.scala-lang.org/`) is the most recent version of the language. We have the option of running Scala code online using our browser or installing and running it on our machine. Running it online is good for getting started with Scala as a first step; however, you will need to install it on your computer to learn the language's in-depth features and make use of it for data analysis.

Running Scala code online

There are some great resources available online to run code in your web browser. These are good for trying out small snippets of Scala code and gaining a better understanding of how Scala works. Some of these online resources allow you to share your code with someone else by generating a static URL that is typically valid for a few months. This feature could be useful for quick code review or collaboration with someone on the internet.

Let's look at two such online resources:

- Scastie
- ScalaFiddle

Scastie

Scastie (`https://scastie.scala-lang.org`) is a great online resource for trying out small Scala code snippets. All that is needed is a web browser and access to the internet.

The main screen is divided into two parts: the top part consists of a program and its output and the bottom part is the output from the backend server that compiles and runs the code. You can modify the code and run it any number of times by using the **Run** option, as shown in the following screenshot:

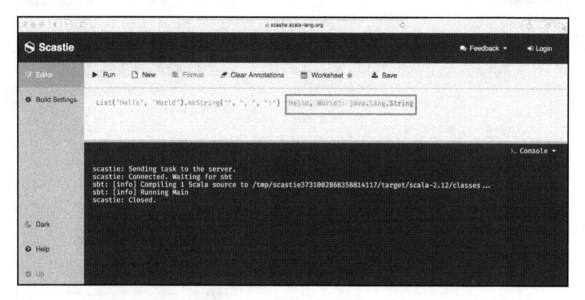

When run in the browser, this simple example displays what is going on in each step. In our example, the Scala expression produced `Hello, World!`, which is of the `java.lang.String` type. The expression is of the following pattern:

List of strings joined together by a comma with an exclamation mark suffix.

ScalaFiddle

ScalaFiddle (`https://scalafiddle.io/`) is another good online resource for running Scala code. This is a good resource for sharing your code with someone else by generating a URL after you save; however, it does require the user to be logged in to a GitHub account. The ScalaFiddle homepage is shown in the following screenshot:

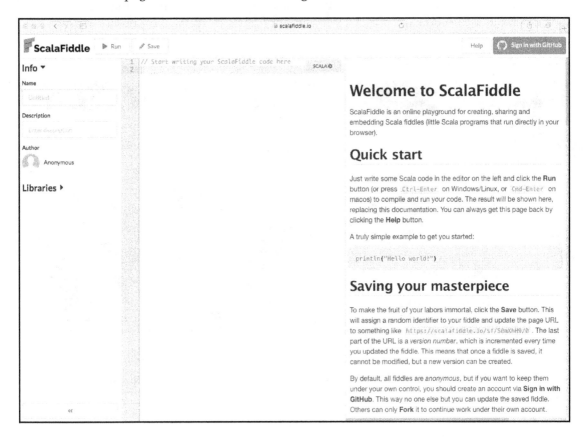

Installing Scala on your computer

An excellent resource for Scala installation on your computer is the Scala official page at `https://www.scala-lang.org/download/`. It mentions how important it is to make sure that Java JDK 8 is installed on your machine. This can be checked by running the following command:

```
$ java -version
```

If it is missing or if you have an older version of JDK, please follow the instructions on Java SE downloads
from `https://www.oracle.com/technetwork/java/javase/downloads/index.html` for your OS. Open JDK (`http://openjdk.java.net/install/`) is another resource for installing JDK.

 Please note that a higher version of Java might work fine as well; however, the examples provided in this book have only been tested on Java 8.

A successfully installed JDK 8 should output something similar to the following code:

```
$ java -version
java version "1.8.0_181"
Java(TM) SE Runtime Environment (build 1.8.0_181-b13)
Java HotSpot(TM) 64-Bit Server VM (build 25.181-b13, mixed mode)

$ javac -version
javac 1.8.0_181
```

Installing command-line tools

The **Scala Build Tool** (**SBT**) is a command-line tool that is very popular for building Scala projects. It also provides a Scala console that can be used for exploring Scala language features and its API.

The following are the SBT installation instructions for macOS using the **Homebrew** tool. The SBT installation will vary from one OS to the other. For more details on SBT, please refer to the official SBT page at `https://www.scala-sbt.org/index.html`:

1. Install Homebrew first, if it is not already installed:

   ```
   $ /usr/bin/ruby -e "$(curl -fsSL
   https://raw.githubusercontent.com/Homebrew/install/master/install)"
   # Install homebrew if not already installed
   ```

2. Install `sbt` using Homebrew, as follows:

   ```
   $ brew install sbt@1
   ```

To verify that SBT and Scala are installed correctly on your machine, go through the following steps:

1. Run the `sbt` command and then run the `console` command inside `sbt` to get access to the Scala console, as follows:

```
$ sbt
[info] Loading project definition from /Users/handsonscala/project
[info] Set current project to handsonscala (in build
                                file:/Users/handsonscala/)
[info] sbt server started at
local:///Users/handsonscala/.sbt/1.0/server/b6ef035291e7ae427145
/sock

sbt:handsonscala> console
[info] Starting scala interpreter...
Welcome to Scala 2.12.6 (Java HotSpot(TM) 64-Bit Server VM, Java
1.8.0_181).
Type in expressions for evaluation. Or try :help.

scala>
```

2. Run `:quit` to exit the Scala console. To exit `sbt`, run the `exit` command:

```
scala> :quit

[success] Total time: 6 s, completed Sep 16, 2018 11:29:24 AM
sbt:handsonscala> exit
[info] shutting down server
$
```

Explore Scala from SBT by performing some of the popular Scala `List` operations. To do this, go through the following steps:

1. Start `sbt` and get access to the Scala console, as follows:

```
$ sbt
[info] Loading project definition from /Users/handonscala/project
[info] Set current project to handsonscala (in build
file:/Users/handsonscala/)
[info] sbt server started at
local:///Users/handsonscala/.sbt/1.0/server/b6ef035291e7ae427145/so
ck
sbt:> console
[info] Starting scala interpreter...
Welcome to Scala 2.12.6 (Java HotSpot(TM) 64-Bit Server VM, Java
1.8.0_181).
```

```
Type in expressions for evaluation. Or try :help.

scala>
```

2. Create a Scala `List` of US states using the following code:

```
scala> val someStates = List("NJ", "CA", "IN", "MA", "NY", "AZ",
                             "PA")
someStates: List[String] = List(NJ, CA, IN, MA, NY, AZ, PA)
```

3. Examine the size of the `List` as follows:

```
scala> someStates.size
res0: Int = 7
```

4. Sort the `List` in ascending order as follows:

```
scala> someStates.sorted
res1: List[String] = List(AZ, CA, IN, MA, NJ, NY, PA)
```

5. Reverse the `List` as follows:

```
scala> someStates.reverse
res2: List[String] = List(PA, AZ, NY, MA, IN, CA, NJ)
```

6. Join the elements of the list using a comma (,) as a separator, as shown in the following code:

```
scala> someStates.mkString(",")
res3: String = NJ,CA,IN,MA,NY,AZ,PA
```

7. Perform sort and join operations as a chain, as shown in the following code:

```
scala> someStates.sorted.mkString(",")
res4: String = AZ,CA,IN,MA,NJ,NY,PA
```

8. Finally, exit the Scala console and quit `sbt` once you are done exploring, as shown in the following code:

```
scala> :quit

[success] Total time: 17 s, completed Sep 16, 2018 11:22:41 AM
sbt:someuser> exit
[info] shutting down server
```

Scala's `List` is a powerful data structure, and we will be looking at this and several other commonly used Scala data structures more detail in the later chapters. Scala `List` provides a comprehensive and intuitive API that makes it very easy to work with lists. We previously explored how to construct a Scala `List` and got an idea of some of the commonly used List APIs. The primary objective of this exercise was to make sure that the Scala command-line tools were set up and working correctly on the local computer.

Installing IDE

There are several **integrated development environment** (**IDE**) tools that have support for Scala. JetBrains IntelliJ IDEA Community Edition is a great choice and has excellent Scala support.

Download and install JetBrains IntelliJ IDEA Community Edition (`https://www.jetbrains.com/idea/download/`) for your OS. Once the installation is complete, please follow these steps:

1. Start IntelliJ IDEA Community Edition and accept all default suggested settings. It is always possible to back and change settings at any later point in time. Once this is done, you will see a screen similar to the following screenshot:

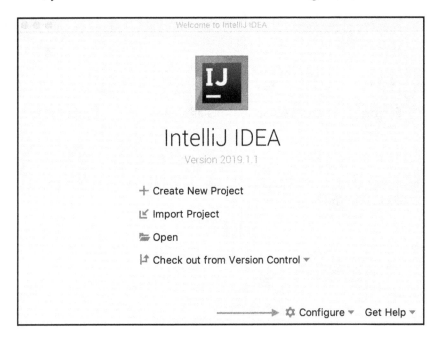

2. Select **Configure** from the preceding screen (highlighted by red color arrow) and then select **Plugins** from the drop-down list. You will see a screen similar to the following screenshot:

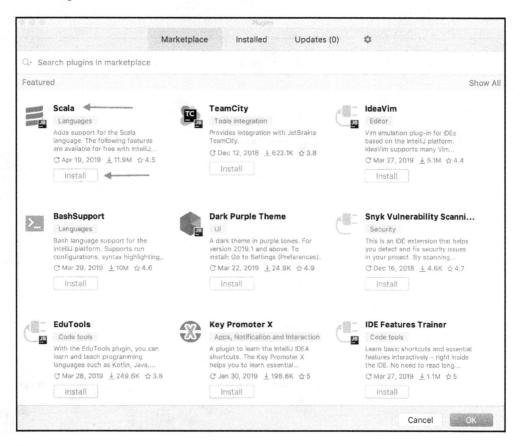

3. Install the Scala plugin by clicking **Install** from the screen aforementioned (highlighted by red color arrow). This action will download the Scala plugin and install it. Once this is completed, you will see a screen similar to the following screenshot:

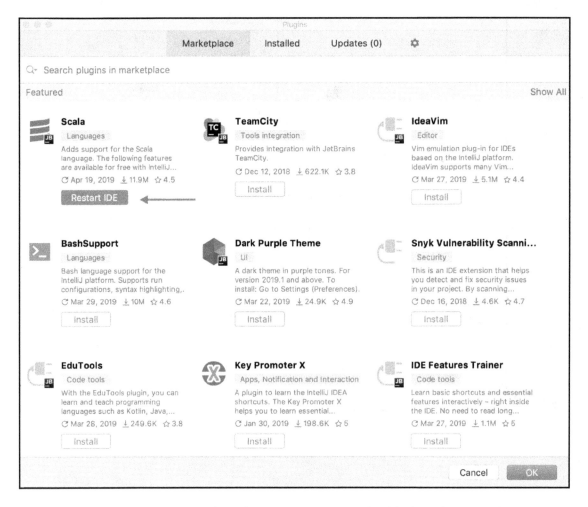

4. Now restart your IDE by selecting **Restart IDE** (highlighted by red color arrow). This will restart your IDE and make it ready for developing Scala code. Now you are ready to start using the IDE for writing Scala code.

Now let us create a sample Scala/SBT project called `HandsOnScala` by following the instructions
at `https://docs.scala-lang.org/getting-started-intellij-track/building-a-scala-project-with-intellij-and-sbt.html`.

The following steps show how to create a new Scala-based SBT project:

1. Click on **Create New Project** as shown in the following screenshot:

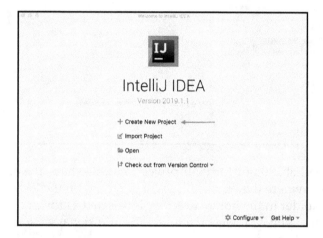

2. Select on **Scala** and then click **Next** to move onto next step, as shown in the following screenshot:

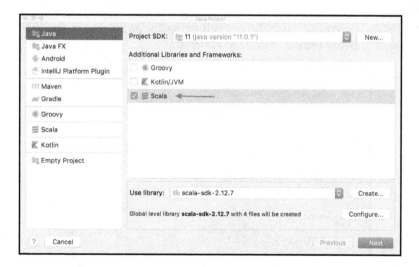

3. Make sure to select **sbt** as highlighted in the following screenshot:

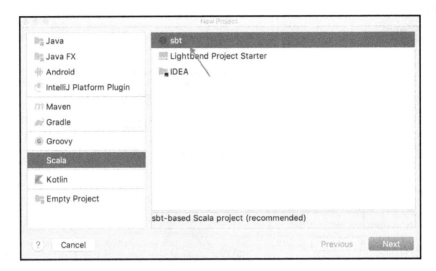

4. Specify a suitable name for your project. You also have the option of specifying the location where the project files will be stored on the local computer. By default, a folder in the home directory is selected as the project location. Please also note that you can select the following additional version details:

   - JDK version
   - SBT version
   - Scala version

   In most cases, you can use the default versions, as shown in the following screenshot, press **Finish** button to continue:

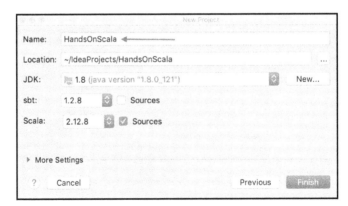

5. Add Scala Framework Support to the project (if it has not already been selected), as shown in the following screenshot. This is very important because, without Scala Framework Support, the IDE will not treat this project as a Scala project and will prevent any Scala-related work in the IDE:

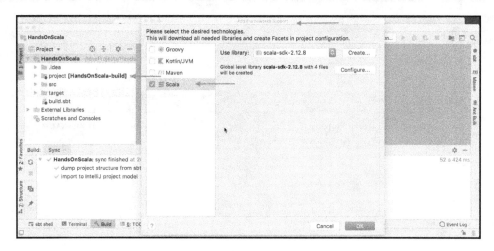

6. Just like in Java, the Scala source code is organized into packages. Create a package called `handson.example` by expanding to `src/main/scala` and right-clicking on the folder shown in the following screenshot:

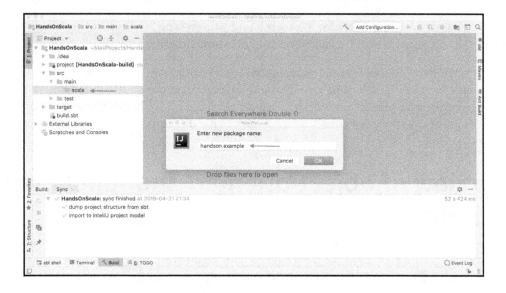

7. Now you are ready to start writing Scala source code and can run it on your IDE. Create and run your first Scala worksheet called `HandsOnScratchPad`, as shown in the following screenshot:

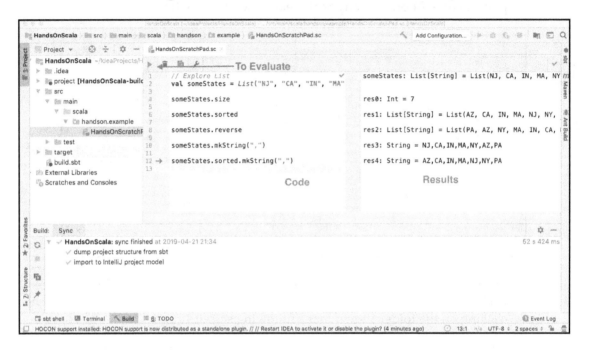

Scala worksheets are great tools for exploring and understanding Scala APIs. You can conveniently see the code and its associated evaluation side by side.

With the preceding IDE set up and running, you are now ready to work on any Scala project using the IntelliJ IDE. You also have the option of conveniently exploring small code snippets of Scala using the Scala worksheet feature of the IDE. This IDE also comes with a debugger and many more features that help improve a developer's productivity.

Overview of object-oriented and functional programming

Scala supports object-oriented and functional styles of programming. Both of these programming paradigms have been proven to help model and solve real-world problems. In this section, we will explore both of these styles of programming using Scala.

Object-oriented programming using Scala

In the object-oriented paradigm, you think in terms of objects and classes. A class can be thought of as a template that acts as a basis for creating objects of that type. For example, a Vehicle class can represent real-world automobiles with the following attributes:

- vin (a unique vehicle identification number)
- manufacturer
- model
- modelYear
- finalAssemblyCountry

A concrete instance of Vehicle, representing a real-world vehicle, could be:

- vin: WAUZZZ8K6AA123456
- manufacturer: Audi
- model: A4
- modelYear: 2009
- finalAssemblyCountry: Germany

Let's put these attributes in action in Scala.

Go to the Scala/SBT console and write the following lines of code:

1. Define Vehicle Scala class as per the preceding specifications:

```
scala> class Vehicle(vin: String, manufacturer: String, model:
                     String,
                     modelYear: Int, finalAssemblyCountry: String)
defined class Vehicle
```

2. Create an instance of Vehicle class:

```
scala> val theAuto = new Vehicle("WAUZZZ8K6AA123456", "Audi", "A4",
                                 2009, "Germany")
theAuto: Vehicle = Vehicle@7c6c2822
```

Following is the IntelliJ Scala worksheet:

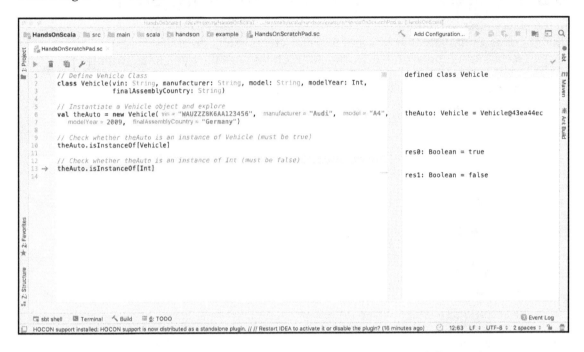

The object-oriented approach puts data and behavior together. The following are core tenets of object-oriented programming:

- **Encapsulation**: It provides a mechanism to shield implementation details and internal properties
- **Abstraction**: It provides constructs such as classes and objects to model real-world problems
- **Inheritance**: It provides constructs to reuse implementation and behavior using subclassing
- **Polymorphism**: It facilitates mechanisms for an object to react to a message based on its actual type

Let's look at encapsulation and abstraction in Scala **Read-Evaluate-Print-Loop**
(**REPL**). We'll use Scala's `construct` class to define a template for a real-world `Vehicle`, as
shown in the following code:

1. Let us define `Vehicle` class, this is an example of abstraction because we are
   taking real-world complex entities and defining a simple model to represent
   them. When internals of implementations is hidden then it is an example of
   encapsulation. Publicly visible methods define behavior:

```scala
scala> class Vehicle(vin: String, manufacturer: String, model:
String, modelYear: Int, finalAssemblyCountry: String) { // class is
an example of abstraction
     | private val createTs = System.currentTimeMillis() // example
of encapsulation (hiding internals)
     | def start(): Unit = { println("Starting...") } // behavior
     | def stop(): Unit = { println("Stopping...") } // behavior
     | }
defined class Vehicle
```

2. Now let create an instance of `Vehicle`. This is also an abstraction because
   `Vehicle` class is a template representing a simplified model of real-world
   vehicles. An instance of `Vehicle` represents a very specific vehicle but it is still a
   model:

```scala
scala> val theAuto = new Vehicle("WAUZZZ8K6AA123456", "Audi", "A4",
     2009, "Germany") // object creation is an example of
abstraction
theAuto: Vehicle = Vehicle@2688b2be
```

3. Perform `start` behavior on the object:

```scala
scala> theAuto.start()
Starting...
```

4. Perform `stop` behavior on the object:

```scala
scala> theAuto.stop()
Stopping...
```

To reiterate the main points aforementioned, the ability to define a class is an example of
abstraction. Inside the class, we have an attribute called `createTs` (creation timestamp).
The scope of this attribute is private and this attribute cannot be accessed from outside the
class. The ability to hide internal details is an example of **encapsulation**.

Now let's look at inheritance and polymorphism in Scala REPL. We'll define a new class called `SportsUtilityVehicle` by extending the `Vehicle` class, as shown in the following code:

1. Define `SportsUtilityVehicle` class that provides an extension to `Vehicle` class:

```
scala> class SportsUtilityVehicle(vin: String, manufacturer:
String, model: String, modelYear: Int, finalAssemblyCountry:
String, fourWheelDrive: Boolean) extends Vehicle(vin, manufacturer,
model, modelYear, finalAssemblyCountry) { // inheritance example
    | def enableFourWheelDrive(): Unit = { if (fourWheelDrive)
println("Enabling 4 wheel drive") }
    | override def start(): Unit = {
    | enableFourWheelDrive()
    | println("Starting SUV...")
    | }
    | }
defined class SportsUtilityVehicle
```

2. Create an instance of `SUV` object but assign to `Vehicle` type object, this is permissible because every SUV object is also a `Vehicle`:

```
scala> val anotherAuto: Vehicle = new
SportsUtilityVehicle("WAUZZZ8K6A654321", "Audi", "Q7", 2019,
                      "Germany", true)
anotherAuto: Vehicle = SportsUtilityVehicle@3c2406dd
```

3. Perform start behavior on the object, on doing so the object exhibits the behavior of an SUV class. This is the polymorphism property facilitated by the object-oriented paradigm:

```
scala> anotherAuto.start() // polymorphism example
Enabling 4 wheel drive
Starting SUV...
```

Inheritance is a powerful construct that allows us to reuse code. We created an instance of `SportsUtilityVehicle` and assigned it to a type of vehicle. When we invoke the `start` method on this object, the runtime system automatically determines the actual type of object and calls the `start` method defined in `SportsUtilityVehicle`. This is an example of polymorphism, where we can treat objects as base types; however, at runtime, the appropriate behavior is applied depending upon the true type of the object.

The following is a UML diagram with a more formal representation of the inheritance relationship:

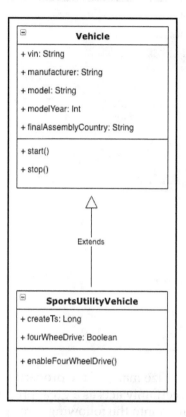

It captures the following important properties:

- The **Vehicle** is a super-class or base-class
- **SportsUtilityVehicle** is a sub-class that extends the **Vehicle** base-class
- This relationship can be envisioned as a parent-child relationship

This diagram shows that **Vehicle** is a base class and the SportsUtilityVehicle subclass extends this base class. The subclass adds its own additional attributes and behavior. One way to look at inheritance is as a generalization-specialization construct. The base class represents a generalized set of attributes and behavior. The derived class or subclass adds its specialization by either altering some of the base class behaviors or adding its own.

The following is a screenshot of the same example in IDE:

IDEs such IntelliJ help us to visualize many of the properties of classes and objects in a user-friendly way. A good IDE certainly acts as a great productivity tool. In the preceding example, the IDE screen is divided into the following three parts:

- **Structure**: Structural properties of classes and objects, such as methods and attributes
- **Source code**: Source code in the context
- **Runtime**: Output from the execution of the program

Functional programming using Scala

In the functional programming paradigm, functions become the primary tool for modeling solutions to a problem. In the simplest form, we can think of a function as something that accepts one or more input and produces an output.

To illustrate this concept, let's define a function in Scala that accepts two sets of integer input and returns the sum of two integers plus one, as shown in the following code:

```scala
scala> val addAndInc = (a: Int, b: Int) => a + b + 1
addAndInc: (Int, Int) => Int = <function2>

scala> addAndInc(5, 10)
res0: Int = 16
```

In the preceding example, we have created an anonymous function that takes two sets of integer input and returns an integer output. This function increments the sum of two input numbers and returns the result. There is another way of defining a function in Scala as a named method. Let's look at that in the Scala REPL:

```scala
scala> def addAndIncMethod(a: Int, b: Int) = a + b + 1
addAndIncMethod: (a: Int, b: Int)Int

scala> addAndIncMethod(5, 10)
res1: Int = 16

scala> val methodAsFunc = addAndIncMethod
<console>:12: error: missing argument list for method addAndIncMethod
Unapplied methods are only converted to functions when a function type is
expected.
You can make this conversion explicit by writing `addAndIncMethod _` or
`addAndIncMethod(_,_)` instead of `addAndIncMethod`.
       val methodAsFunc = addAndIncMethod
```

In the preceding example, we have defined a method that is bound to a name. The usage of the anonymous function and the named method is identical; however, there are some subtle differences between the two, as shown in the following list:

- The signature of the anonymous function is `(Int, Int) => Int` and the signature of the named method is `(a: Int, b: Int)Int`
- When we try to assign a method to a variable, we get an error

A method can be easily converted into an anonymous function in Scala by doing the following:

```scala
scala> val methodAsFunc = addAndIncMethod _ // turns method into function
methodAsFunc: (Int, Int) => Int = <function2>

scala> methodAsFunc(5, 10)
res2: Int = 16
```

As can be seen, after conversion, the signature changes to `(Int, Int) => Int`.

Anonymous functions and named methods are both useful in the context of Scala programming. In the previous section on Scala's object-oriented programming, we defined a `scala` class with some methods. These were all named methods, and would not be of much value from an object-oriented point of view if these were anonymous methods.

The object-oriented approach puts data and behavior together. Objects have mutable states that are manipulated by methods operating on those objects. From a purely functional approach, inputs are never mutated and new outputs are created when a function is applied to these inputs. There is a strong emphasis on immutability in the functional paradigm. Immutability has strong advantages when it comes to the reasoning behind the behavior of a program. You do not have to be concerned about the accidental corruption of a shared state as there are no shared states in the pure functional paradigm.

Functions can be defined within a function. Let's look at the following concrete example using Scala REPL to see why this is useful:

```
scala> def factorial(n: Int): Long = if (n <= 1) 1 else n * factorial(n-1)
factorial: (n: Int)Int

scala> factorial(5)
res0: Long = 120
```

The preceding code is an example of the `factorial` function, and it uses recursion to compute this value. We know that classic recursion requires a stack size proportional to the number of iterations. Scala provides tail recursion optimization to address this problem, and we can make use of an inner function to optimize this problem. In the following code, we'll define a function inside a function:

```
scala> import scala.annotation.tailrec
import scala.annotation.tailrec

scala> def optimizedFactorial(n: Int): Long = {
     |   @tailrec
     |   def factorialAcc(acc: Long, i: Int): Long = {
     |   if (i <= 1) acc else factorialAcc(acc * i, i -1)
     |   }
     |   factorialAcc(1, n)
     |   }
optimizedFactorial: (n: Int)Long

scala> optimizedFactorial(5)
res1: Long = 120
```

We can call the `factorialAcc` function an inner function and the main function, `optimizedFactorial`, an outer function. The outer function preserves the same interface as the factorial function defined earlier. The inner function uses an accumulator design pattern that allows tail recursion optimization to work. In each iteration, the results are accumulated into an accumulator variable that is not dependent on the iteration variable.

Recursion is a very useful programming construct and tail recursion is a special type of recursion that can be optimized at compile time. Let us first try to understand what recursion really is and what type of recursion is considered tail recursive. In simple terms, any function that calls or invokes itself one or more times is considered recursive. Our factorial examples in both forms are recursive. Let us look at the execution of the first version for a value of 5:

```
factorial(5)
5 * factorial(4)
5 * 4 * factorial(3)
5 * 4 * 3 * factorial(2)
5 * 4 * 3 * 2 * factorial(1)
5 * 4 * 3 * 2 * 1
120
```

For the second version:

```
optimizedFactorial(5)
factorialAcc(1, 5)
factorialAcc(1*5, 4)
factorialAcc(1*5*4, 3)
factorialAcc(1*5*4*3, 2)
factorialAcc(1*5*4*3*2, 1)
1*5*4*3*2
120
```

The most important difference between the two versions is in the last return statement of the function:

```
// output of next invocation must be multiplied to current number, so //
the state (current number) has to preserved on stack frame
n * factorial(n-1)
```

```
// function being called already knows the current state being passed // as
the first argument so it does not need preserved on stack frame
factorialAcc(acc * i, i -1)
```

Recursive functions become expensive due to the usage of stack frames proportional to the number times a function invokes itself. Just imagine running factorial for a large number, it would easily cause stack overflow to occur fairly quickly. Tail recursion property of a recursive algorithm allows us to perform some optimizations to reduce the stack usage as if it was an iterative algorithm.

The following is a screenshot of the preceding recursion examples in IntelliJ IDE. The IDE helps us clearly see which functions or methods are purely **Recursive** and which ones are **Tail Recursive**:

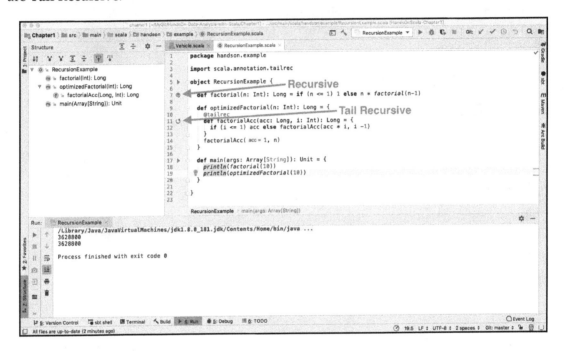

Please note the specific symbols next to `factorial` and `optimizedFactorial`. The two symbols are different, and if you hover over them, you can see the full description, listed as follows:

- Method factorial is recursive
- Method factorial is tail recursive

Let's use the following code to see whether we are able to apply tail recursion optimization to the original factorial function in Scala REPL:

```
scala> @tailrec
     | def factorial(n: Int): Long = if (n <= 1) 1 else n * factorial(n-1)
<console>:14: error: could not optimize @tailrec annotated method
```

```
factorial: it contains a recursive call not in tail position
        def factorial(n: Int): Long = if (n <= 1) 1 else n * factorial(n-1)
```

As you can see from the error message, tail recursion optimization cannot be applied in this case. The Scala compiler is able to perform the tail recursion optimization of a recursive function when it follows certain constraints. The details of this are not within the scope of this book; however, this is a very important and useful feature provided by Scala. As we saw in the preceding code, by using an inner function and an accumulator design pattern, we are able to successfully achieve this optimization. As a result of this optimization, the stack usage requirements for the recursive function become equivalent to that of the iterative function.

Functional programming treats functions as first-class citizens. In fact, Scala's collection API heavily makes use of this feature. Functions can be defined within a function and these can be passed as parameters.

From a data analysis and processing perspective, functional programming provides a framework for massively parallel processing solutions. If the input data is immutable, transformations can be applied to the input any number of times and will always result in the same output.

Next, we will look at Scala's case classes and the collection API, which provide some major advantages in data analysis.

Scala case classes and the collection API

Scala case classes and its collection API play a significant role in data analysis using Scala. This section will give you insight into these topics and an understanding of their relevance in the context of data analysis.

During the data analysis process, we will frequently encounter data that consists of a collection of records. These records often need to be transformed, cleaned, or filtered.

Scala case classes

Scala case classes provide a convenient mechanism to work with objects that hold values. Let's look at an example in Scala REPL. The case class defined in the following code will be used in other example codes in this chapter:

```
scala> case class Person(fname: String, lname: String, age: Int)
defined class Person
```

```
scala> val jon = Person("Jon", "Doe", 21)
jon: Person = Person(Jon,Doe,21)
```

In the preceding example, we have defined a Scala case class called `Person` with three attributes, namely `fname`, `lname`, and `age`. We created an instance, `jon`, of the `Person` class without using the new keyword. Also, note that the `jon` object's attributes are printed out in a easy-to-use form. There are several such convenient features associated with Scala case classes that are extremely beneficial for programmers in general, particularly someone who deals with data.

Let's look at another convenient feature of Scala case classes, namely the `copy` object. We'll copy a Scala `case` class object by updating only the `fname` attribute using the following code:

```
scala> case class Person(fname: String, lname: String, age: Int)
defined class Person

scala> val jon = Person("Jon", "Doe", 21)
jon: Person = Person(Jon,Doe,21)

scala> val jonNew = jon.copy(fname="John")
jonNew: Person = Person(John,Doe,21)
```

This feature comes in really handy during data processing when we work with a template representation and generate specific instances from a template by updating a subset of attributes.

Another great feature of case classes is pattern matching, which helps in writing flexible code that is easier to work with. Let's look at an example of pattern matching in Scala REPL, as shown in the following code:

```
scala> def isJon(p: Person) = {
     | p match {
     | case Person("Jon", _, _) => {println("I am Jon"); true}
     | case Person(n,_,_) => {println(s"I am not Jon but I am ${n}");
false}
     | case _ => {println("I am not Jon but I am something other than
Person"); false}
     | }
     | }
isJon: (p: Person)Boolean

scala> val jon = Person("Jon", "Doe", 25)
jon: Person = Person(Jon,Doe,25)

scala> isJon(jon)
```

```
I am Jon
res13: Boolean = true

scala> val bob = Person("Bob", "Crew", 27)
bob: Person = Person(Bob,Crew,27)

scala> isJon(bob)
I am not Jon but I am Bob
res14: Boolean = false

scala> isJon(null)
I am not Jon but I am something other than Person
res16: Boolean = false
```

We can explore the same example in the IDE, as shown in the following screenshot:

Using the IDE, we can clearly see the properties of the `case` class. Another great option is to use the Scala worksheet feature in IDE to explore this example, as shown in the following screenshot:

The preceding screenshot shows a fairly simple example of pattern matching using Scala case classes that illustrates the simplicity and power of this feature. In the data analysis world, pattern matching has been found to be extremely useful in solving certain categories of problems. Scala provides an intuitive, elegant, and simple way to take advantage of pattern matching.

Let's look at the preceding example in a bit more detail, looking at the following lines:

- **Line #4**: `case Person("Jon", _, _)` means any person whose first name is Jon
- **Line #7**: `case Person(n, _, _)` means any person with the first name is extracted into variable `n`
- **Line #10**: `case _` means anything that does not match line #4 and line #7

With classic pattern matching, it is generally necessary for you to write a significant amount of boilerplate code with `if-then-else` types of constructs. Scala and its case classes provide a concise and expressive way to solve this problem.

Scala collection API

Scala has a comprehensive API for working conveniently with collections. A good understanding of some of the APIs is essential for making effective use of Scala in data analysis.

Although a full review of the Scala collection API is not part of the scope of this book, three key data structures will be covered in this section: the array, list, and map. The emphasis here is on their direct relevance to data analysis. For complete details, please refer to the official Scala resource at `https://www.scala-lang.org/api/current/scala/collection/index.html`.

It is also important to consider the performance characteristics of a Scala collection API and use this information in making appropriate data structure selections for the problem being solved. Refer to `https://docs.scala-lang.org/overviews/collections/performance-characteristics.html` for more information.

Array

The array provides fast and constant time performance for the random access (index-based access) of a collection. Arrays can be thought of as a data structures that are backed by an array of bytes that are contiguously laid out in the computer's memory. This means that individual elements of an array are placed one after the other in memory. With a layout like this, there is an implicit connection between the elements and the physical position of the elements in the memory, determining which one is first, next, and so on. For example, the position of the tenth element in an array can be determined by adding the sizes of the first nine elements.

The following is an example run execution in Scala REPL to demonstrate the Scala array and some of its functionality:

```
scala> val persons = Array(Person("Jon", "Doe", 21), Person("Alice",
"Smith", 20), Person("Bob", "Crew", 27)) // construct a Array of Person
objects
persons: Array[Person] = Array(Person(Jon,Doe,21), Person(Alice,Smith,20),
Person(Bob,Crew,27))

scala> val personHead = persons.head // first person in the collection
personHead: Person = Person(Jon,Doe,21)

scala> val personAtTwo = persons(2) // person at index 2 (this is same as
apply operation)
personAtTwo: Person = Person(Bob,Crew,27)
```

```
scala> val personsTail = persons.tail // collection without the first
person
personsTail: Array[Person] = Array(Person(Alice,Smith,20),
Person(Bob,Crew,27))

scala> val personsByAge = persons.sortBy(p => p.age) // sort persons by age
personsByAge: Array[Person] = Array(Person(Alice,Smith,20),
Person(Jon,Doe,21), Person(Bob,Crew,27))

scala> val personsByFname = persons.sortBy(p => p.fname) // sort persons by
first name
personsByFname: Array[Person] = Array(Person(Alice,Smith,20),
Person(Bob,Crew,27), Person(Jon,Doe,21))

scala> val (below25, above25) = persons.partition(p => p.age <= 25) //
split persons by age
below25: Array[Person] = Array(Person(Jon,Doe,21), Person(Alice,Smith,20))
above25: Array[Person] = Array(Person(Bob,Crew,27))

scala> val updatePersons = persons.updated(0, Person("Jon", "Doe", 20)) //
update first element
updatePersons: Array[Person] = Array(Person(Jon,Doe,20),
Person(Alice,Smith,20), Person(Bob,Crew,27))
```

The following is a summary of the array operations and their associated performance characteristics:

Array operation	Purpose	Performance
head	Get the element at the head	O(1) constant time
tail	Get elements other than the head	O(n) linear time, proportional to the number of elements in the collection
apply	Get the element at a specified index	O(1) constant time
update	Replace the element at the specified index	O(1) constant time
prepend	Add a new element at the head	Not supported for an array
append	Add a new element at the end	Not supported for an array
insert	Insert a new element at the specified index	Not supported for an array

As can be seen in the preceding table, the `apply` operation for getting the element at a specified index is a fast constant-time operation for an array. Along similar lines, the `update` operation for replacing an element at the specified index is also a fast constant-time operation. On the other hand, the `tail` operation for getting elements other than the `head` is a slow linear time operation. In fact, the `prepend`, `append`, and `insert` operations are not even supported for an array. This might seem a limiting factor at first, but Scala has an `ArrayBuffer` class for building an array, and that should be used if such operations are necessary.

In data analysis, we typically create a dataset initially and use it over and over again during different phases of the analysis. This implies that the dataset construction is generally a one-time step, and the construction is then used multiple times. This is precisely why a builder such as `ArrayBuffer` is separated from the array: because each serves a different purpose. The `ArrayBuffer` is designed to help in the construction of the array with support for the commonly desired build operations. Let's look at the `ArrayBuffer` functionality using Scala REPL, as shown in the following code:

```
scala> import scala.collection.mutable.ArrayBuffer
import scala.collection.mutable.ArrayBuffer

scala> val personsBuf = ArrayBuffer[Person]() // create ArrayBuffer of
Person
personsBuf: scala.collection.mutable.ArrayBuffer[Person] = ArrayBuffer()
scala> personsBuf.append(Person("Jon", "Doe", 21)) // append a Person
object at the end
scala> personsBuf.prepend(Person("Alice", "Smith", 20)) // prepend a Person
object at head
scala> personsBuf.insert(1, Person("Bob", "Crew", 27)) // insert a Person
object at index 1
scala> val persons = personsBuf.toArray // materialize into an Array of
Person
persons: Array[Person] = Array(Person(Alice,Smith,20), Person(Bob,Crew,27),
Person(Jon,Doe,21))
scala> val personRemoved = personsBuf.remove(1) // remove Person object at
index 1
personRemoved: Person = Person(Bob,Crew,27)

scala> val personsUpdated = personsBuf.toArray // materialize into an Array
of Person
personsUpdated: Array[Person] = Array(Person(Alice,Smith,20),
Person(Jon,Doe,21))
```

As can be seen in the preceding code, `ArrayBuffer` provides a comprehensive set of functionalities to construct a collection and provides a convenient mechanism to materialize it into an array once construction is complete.

List

A `List` provides fast and constant time performance for `head` and `tail` operations in a collection. We can visualize a List as a collection of elements that are connected by some kind of link. Let's look at the Scala `List` functionality using Scala REPL, as shown in the following code:

```
scala> val persons = List(Person("Jon", "Doe", 21), Person("Alice",
"Smith", 20), Person("Bob", "Crew", 27)) // construct a List of Person
objects
persons: List[Person] = List(Person(Jon,Doe,21), Person(Alice,Smith,20),
Person(Bob,Crew,27))

scala> val personHead = persons.head // first person in the collection
personHead: Person = Person(Jon,Doe,21)

scala> val personAtTwo = persons(2) // person at index 2 (this is same as
apply operation)
personAtTwo: Person = Person(Bob,Crew,27)

scala> val personsTail = persons.tail // collection without the first
person
personsTail: List[Person] = List(Person(Alice,Smith,20),
Person(Bob,Crew,27))

scala> val personsByAge = persons.sortBy(p => p.age) // sort persons by age
personsByAge: List[Person] = List(Person(Alice,Smith,20),
Person(Jon,Doe,21), Person(Bob,Crew,27))

scala> val personsByFname = persons.sortBy(p => p.fname) // sort persons by
first name
personsByFname: List[Person] = List(Person(Alice,Smith,20),
Person(Bob,Crew,27), Person(Jon,Doe,21))

scala> val (below25, above25) = persons.partition(p => p.age <= 25) //
split persons by age
below25: List[Person] = List(Person(Jon,Doe,21), Person(Alice,Smith,20))
above25: List[Person] = List(Person(Bob,Crew,27))

scala> val updatePersons = persons.updated(0, Person("Jon", "Doe", 20)) //
update first element
updatePersons: List[Person] = List(Person(Jon,Doe,20),
Person(Alice,Smith,20), Person(Bob,Crew,27))
```

The following is a summary of the List operations and their associated performance characteristics:

List operation	Purpose	Performance
head	Get the element at the head	O(1) constant time
tail	Get elements other than the head	O(1) constant time
apply	Get the element at the specified index	O(n) linear time, proportional to the number of elements in the collection
update	Replace the element at the specified index	O(n) linear time, proportional to the number of elements in the collection
prepend	Add a new element at the head	O(1) constant time
append	Add a new element at the end	O(n) linear time, proportional to the number of elements in the collection
insert	Insert a new element at the specified index	Not supported for list

As can be seen in the preceding table, the List enables very fast head, tail, and prepend operations. For the array type described earlier, we saw that tail was an expensive linear time operation. The apply operation for getting an element at the specified index is a linear time operation. This is because the desired element can only be located by traversing the links, starting from the head. This explains why an update is a slow operation for the List.

In a real-world scenario, constant time performance is the desired behavior and we want to avoid linear time performance, particularly for large datasets. Performance is an important factor in determining the most suitable data structure for the problem being solved. If constant time performance is not practical, we generally look for data structures and algorithms that provide *Log Time performance O(log n)*: time proportional to the logarithm of the collection size. Note that there are many algorithms, such as sorting, with best performance times of *O(n log n)*. When dealing with large datasets, a good understanding of the performance characteristics of the data structures and algorithms that are used goes a long way in solving problems effectively and efficiently.

Similar considerations hold true for memory usage, even though larger amounts of RAM are now becoming available at a cheaper price. This is because the growth in the size of data being produced is much higher than the drop in prices of RAM.

Let's now look at `ListBuffer`, which can be used for constructing a list more efficiently. This will be very useful, given that the `append` operation has significant performance overheads. As mentioned earlier, datasets are generally constructed once but are used multiple times during data analysis processes. Let's look at the following code:

```
scala> import scala.collection.mutable.ListBuffer
import scala.collection.mutable.ListBuffer

scala> val personsBuf = ListBuffer[Person]() // create ListBuffer of Person
personsBuf: scala.collection.mutable.ListBuffer[Person] = ListBuffer()

scala> personsBuf.append(Person("Jon", "Doe", 21)) // append a Person
object at end

scala> personsBuf.prepend(Person("Alice", "Smith", 20)) // prepend a Person
object at head

scala> personsBuf.insert(1, Person("Bob", "Crew", 27)) // insert a Person
object at index 1

scala> val persons = personsBuf.toList // materialize into a List of Person
persons: List[Person] = List(Person(Alice,Smith,20), Person(Bob,Crew,27),
Person(Jon,Doe,21))

scala> val personRemoved = personsBuf.remove(1) // remove Person object at
index 1
personRemoved: Person = Person(Bob,Crew,27)

scala> val personsUpdated = personsBuf.toList // materialize into a List of
Person
personsUpdated: List[Person] = List(Person(Alice,Smith,20),
Person(Jon,Doe,21))
```

If we compare `ArrayBuffer` and `ListBuffer`, we can see that they both offer similar APIs. Their primary use is for constructing an array and list respectively, providing good performance characteristics.

The decision of choosing between array and list is dependent on how the dataset will be used. The following are some useful tips:

- An array should generally be the first choice because of its storage efficiency. Array operations are somewhat limited compared to list operations, and the usage pattern becomes the determining factor.
- If a `tail` operation is necessary, a list is the obvious choice. In fact, there are many recursive algorithms that make extensive use of this feature. Using an array instead of a list will result in a significant performance penalty.

- If `apply` or `update` operations are desired, then an array is certainly a better choice.
- If the `prepend` operation is needed or if a limited use of `append` is required, then a list is the only choice because an array does not support the `prepend` or `append` operations.

As you can see, there are many factors at play when it comes to selecting the appropriate data structure. This is often the case in any software design decision where there are conflicting needs and you need to decide how to make trade-offs. For example, you might decide in favor of using a list even though none of the non-array features of a list are required based on current usage patterns. This could be because of the list's fast `tail` operation, which could be beneficial for the recursive algorithms in future usage patterns.

Recursive algorithms play a central role in functional programming. In fact, Scala supports tail-recursion optimization out of the box, which facilitates practical usage of recursive algorithms with large datasets. With classic recursion, a significant amount of stack space is required, making it impractical to use on large datasets. With tail-recursion optimization, the Scala compiler eliminates the stack growth under the hood. Let's look at a classic recursion and tail-recursion example:

```scala
scala> import annotation.tailrec
import annotation.tailrec

scala> @tailrec def classicFactorial(n: Int): Int = { require(n > 0); if (n
== 1) n else n * classicFactorial(n-1) } // this should fail
<console>:14: error: could not optimize @tailrec annotated method
classicFactorial: it contains a recursive call not in tail position
       @tailrec def classicFactorial(n: Int): Int = { require(n > 0); if (n
== 1) n else n * classicFactorial(n-1) }

scala> def classicFactorial(n: Int): Int = { require(n > 0, "n must be non-
zero and positive"); if (n == 1) n else n * classicFactorial(n-1) } // this
should work
classicFactorial: (n: Int)Int

scala> val classicResult = classicFactorial(5)
classicResult: Int = 120
scala> def tailRecFactorial(n: Int): Int = {
     | require(n > 0, "n must be non-zero and positive")
     | @tailrec def factorial(acc: Int, m: Int): Int = if (m == 1) acc else
factorial(acc * m, m-1) // this should work as this recursive algorithm
meets tail recursion requirements
     | factorial(1, n)
     | }
tailRecFactorial: (n: Int)Int
```

```
scala> val tailRecResult = tailRecFactorial(5)
tailRecResult: Int = 120
```

The preceding examples provides insight into recursive functions, and in particular demonstrates a tail-recursion variant. Let's look at the following example of a tail-recursion algorithm using `List`:

```
scala> val persons = List(Person("Jon", "Doe", 21), Person("Alice",
"Smith", 20), Person("Bob", "Crew", 27))
persons: List[Person] = List(Person(Jon,Doe,21), Person(Alice,Smith,20),
Person(Bob,Crew,27))

scala> @tailrec def minAgePerson(acc: Option[Person], lst: List[Person]):
Option[Person] = {
     | if (lst.isEmpty) acc
     | else if (acc.isEmpty) minAgePerson(Some(lst.head), lst.tail)
     | else if (acc.get.age <= lst.head.age) minAgePerson(acc, lst.tail)
     | else minAgePerson(Some(lst.head), lst.tail)
     | }
minAgePerson: (acc: Option[Person], lst: List[Person])Option[Person]

scala> val youngest = minAgePerson(None, persons) // Person with minimum
age
youngest: Option[Person] = Some(Person(Alice,Smith,20))

scala> val youngestEmpty = minAgePerson(None, Nil) // Nil == List(), an
empty list
youngestEmpty: Option[Person] = None
```

The preceding code is a very simple example of finding a `Person` with the minimum age from a list of `Person` objects. This simple example, however, illustrates the following important and powerful points regarding Scala:

- It is fairly straightforward to write a tail-recursive algorithm using a list in Scala that accumulates information. This algorithm can traverse the entire list without incurring the overhead of stack growth in a classic recursion.
- Scala's `option` construct provides a convenient way of representing the presence or absence of an object.
- List's `head` and `tail` operations come in handy in writing such recursive algorithms, and provide the desired constant time performance for both these operations.
- The code is concise and works even on the empty list.
- Using the accumulator is a commonly used pattern in turning a classic recursion algorithm into a tail-recursion algorithm.

Map

A Map provides a mapping from a key to the associated value. Lookups into a Map based on a key have a generally constant time of *O(1)*. A Map is an important data structure that has many applications in the real world.

Let's look at some simple Map usage using Scala REPL:

```
scala> val countryToCurrency = Map(("US" -> "USD"), ("DE" -> "EUR"), ("FR"
-> "EUR"), ("IN" -> "INR")) // Mapping from country code to currency code
countryToCurrency: scala.collection.immutable.Map[String,String] = Map(US
-> USD, DE -> EUR, FR -> EUR, IN -> INR)

scala> countryToCurrency.keys // country codes
res4: Iterable[String] = Set(US, DE, FR, IN)

scala> countryToCurrency.values // currency codes
res5: Iterable[String] = MapLike.DefaultValuesIterable(USD, EUR, EUR, INR)

scala> countryToCurrency("US") // lookup currency code for US
res6: String = USD
```

In Scala, there are many different types of map, each with its own set of characteristics. We will cover the following three:

Map type	Description	lookup	add	remove	min
HashMap	Backed by a hash table	Near constant time	Near constant time	Near constant time	Linear
TreeMap	Backed by a sorted tree	Log *n*	Log *n*	Log *n*	Log *n*
LinkedHashMap	Insert order is preserved	Linear	Constant time	Linear	Linear

The following are some general considerations to bear in mind regarding the performance of each of these Map types:

- HashMap is the best choice in most cases, particularly for lookup-centric use cases. HashMap does not preserve key insertion order or sort keys.
- TreeMap is suitable for use cases where keys need to be sorted.
- LinkedHashMap is most suited when the key insertion order needs to be preserved.

Let's explore some of these map types in Scala REPL using the following code:

1. Import `HashMap`, `TreeMap`, and `LinkedHashMap` from Scala's
    `collection.mutable` package. Each represents Map type but with a slightly
    different flavor:

    ```scala
    scala> import collection.mutable.{HashMap,TreeMap,LinkedHashMap}
    import collection.mutable.{HashMap, TreeMap, LinkedHashMap}
    ```

2. Create a `HashMap` that maps a number to its English word equivalent. Notice that
    the order of keys is not preserved. The number 8 was at position 8 in our
    constructor however in the object created it is at the first position:

    ```scala
    scala> val numHashMap = HashMap((1->"one"), (2->"two"),
    (3->"three"), (4->"four"), (5->"five"), (6->"six"), (7->"seven"),
    (8->"eight"), (9->"nine")) // keys can be in any order
    numHashMap: scala.collection.mutable.HashMap[Int,String] = Map(8 ->
    eight, 2 -> two, 5 -> five, 4 -> four, 7 -> seven, 1 -> one, 9 ->
    nine, 3 -> three, 6 -> six)
    ```

3. Add a new entry of 0. This got added at the very end however this is just a
    coincidence, it could have been anywhere:

    ```scala
    // add new mapping, keys can be any order
    scala> numHashMap += (0->"zero")
    res5: numHashMap.type = Map(8 -> eight, 2 -> two, 5 -> five, 4 ->
    four, 7 -> seven, 1 -> one, 9 -> nine, 3 -> three, 6 -> six, 0 ->
    zero)
    ```

4. Create a `TreeMap` similar to `HashMap`. Note that the order of keys is preserved. In
    fact, this is due to keys automatically being sorted by `TreeMap`. Our object
    construction had provided the keys in ascending order:

    ```scala
    // keys must be sorted
    scala> val numTreeMap = TreeMap((1->"one"), (2->"two"),
    (3->"three"), (4->"four"), (5->"five"), (6->"six"), (7->"seven"),
    (8->"eight"), (9->"nine"))
    numTreeMap: scala.collection.mutable.TreeMap[Int,String] =
    TreeMap(1 -> one, 2 -> two, 3 -> three, 4 -> four, 5 -> five, 6 ->
    six, 7 -> seven, 8 -> eight, 9 -> nine)
    ```

5. Add a new entry to `TreeMap` with key as 0. This gets added to the beginning because of key sorting in a `TreeMap`:

```
// add a new mapping, keys must get sorted
scala> numTreeMap += (0->"zero")
res6: numTreeMap.type = TreeMap(0 -> zero, 1 -> one, 2 -> two, 3 ->
three, 4 -> four, 5 -> five, 6 -> six, 7 -> seven, 8 -> eight, 9 ->
nine)
```

6. Create a `LinkedHashMap` similar to `HashMap` and `TreeMap`. Note that keys appear exactly as it was specified in the constructor:

```
// order must be preserved
scala> val numLinkedHMap = LinkedHashMap((1->"one"), (2->"two"),
(3->"three"), (4->"four"), (5->"five"), (6->"six"), (7->"seven"),
(8->"eight"), (9->"nine"))
numLinkedHMap: scala.collection.mutable.LinkedHashMap[Int,String] =
Map(1 -> one, 2 -> two, 3 -> three, 4 -> four, 5 -> five, 6 -> six,
7 -> seven, 8 -> eight, 9 -> nine)
```

7. Add new entry to `LinkedHashMap` with key as 0. This gets added at the very end because `LinkedHashMap` preserves the order of key insertion:

```
// this must be the last element
scala> numLinkedHMap += (0->"zero")
res17: numLinkedHMap.type = Map(1 -> one, 2 -> two, 3 -> three, 4
-> four, 5 -> five, 6 -> six, 7 -> seven, 8 -> eight, 9 -> nine, 0
-> zero)
```

Scala offers a good number of choices in terms of Map implementation. You can choose the best option based on the usage pattern. Similar to the design choices between arrays and lists, at times there are trade-offs that need to be considered in deciding the best Map implementation.

Overview of Scala libraries for data analysis

There are a great number of Scala libraries and frameworks that simplify data analysis tasks. There is a lot of innovation happening regarding the simplification of data analysis-related tasks, from simple tasks such as data cleaning, to more advanced tasks such as deep learning. The following sections focus on the most popular data-centric libraries and frameworks that have seamless Scala integration.

Apache Spark

Apache Spark (`https://spark.apache.org/`) is a unified analytics engine for large-scale data processing. Spark provides APIs for batch as well as stream data processing in a distributed computing environment. Spark's API can be broadly divided into the following five categories:

- **Core**: RDD
- **SQL structured**: DataFrames and Datasets
- **Streaming**: Structured streaming and DStreams
- **MLlib**: Machine learning
- **GraphX**: Graph processing

Apache Spark is a very active open source project. New features are added and performance improvements made on a regular basis. Typically, there is a new minor release of Apache Spark every three months with significant performance and feature improvements. At the time of writing, 2.4.0 is the most recent version of Spark.

The following is Spark core's SBT dependency:

```
scalaVersion := "2.11.12"

libraryDependencies += "org.apache.spark" %% "spark-sql" % "2.4.1"
```

Spark version 2.4.0 has introduced support for Scala version 2.12; however, we will be using Scala version 2.11 for exploring Spark's feature sets. Spark will be covered in more detail in the subsequent chapters.

Breeze

Breeze (see `http://www.scalanlp.org/`, `https://github.com/scalanlp/breeze/wiki/Quickstart` and `https://github.com/scalanlp/breeze` for more details) provides a set of libraries containing support for linear algebra, numerical computing, and optimization.

Breeze aims to be a generic, clean, and powerful numerical processing library without sacrificing performance. Breeze is part of the ScalaNLP umbrella project.

For Breeze's SBT dependency, refer to `https://github.com/scalanlp/breeze`.

Breeze-viz

Breeze-viz (see `https://github.com/scalanlp/breeze/wiki/Quickstart` and `https://github.com/scalanlp/breeze/tree/master/viz` for more details) is a visualization library backed by Breeze and JFreeChart.

For its SBT dependency, refer to `https://github.com/scalanlp/breeze`.

DeepLearning

The `DeepLearning.scala` toolkit (`https://deeplearning.thoughtworks.school/`) is a deep learning toolkit for Scala. This library is provided by ThoughtWorks. The library aims to create neural networks using object-oriented as well as functional programming constructs of Scala.

Epic

Epic (`http://www.scalanlp.org/` and `https://github.com/dlwh/epic`) is a high-performance statistical parser and structured prediction library. Similar to Breeze, Epic is also part of the ScalaNLP umbrella project.

Saddle

Saddle (`https://saddle.github.io/`) is a high-performance data-manipulation library with support for array-backed data structures. This library aims to provide R and Python pandas-like feature sets for working with structured data. For more details on using this library, please refer to the documentation at `https://saddle.github.io/doc/index.html`.

Scalalab

Scalalab (`https://github.com/sterglee/scalalab`) is a MATLAB-like scientific computing library. This library aims to provide an efficient scientific programming environment for JVM. The library strives to optimize performance and uses native C/C++ code for numerical computations.

Smile

Smile (`https://haifengl.github.io/smile/`) is a fast and comprehensive **machine learning** (**ML**) system. It aims to provide ML capabilities, keeping speed and ease of use in mind. At the same time, it strives to provide a fairly comprehensive set of toolkits for ML.

Vegas

Vegas (`https://www.vegas-viz.org/`) is a declarative statistical visualization library. This library aims to be a Matplotlib-like library for Scala and Spark. One of the nice things about this library is that it can be easily integrated with Apache Spark. In addition to Apache Spark, it also supports integration with the Apache Flink framework. Apache Flink is another popular open source platform that is optimized for stream-oriented workloads.

Summary

This chapter provided a high-level overview of the Scala programming language. We looked at some of the object-oriented and functional programming aspects of Scala using applied examples. This chapter touched upon the array, list, and map functionalities of the Scala collection API. These data structures have numerous uses in the data analysis life cycle. The chapter also provided the necessary information to set up and install the Scala tools that are essential for understanding and applying the topics covered in subsequent chapters. Finally, a quick overview of the data-centric Scala libraries was provided. We will be making use of these libraries in the next few chapters to solve specific data life cycle problems.

In the next chapter, we will look at the data analysis life cycle and associated tasks.

Data Analysis Life Cycle

2

Data is ubiquitous, and is increasingly playing a significant role in every aspect of our lives, from helping us to make informed decisions to building intelligent systems.

Data is created when a computer system captures and records some kind of observation or activity; this is true of any example of data being electronically recorded, from something as simple as an **Internet of Things (IoT)** device recording temperature at a given point in time at a specific location to something more complex, such as a human being completing an online purchase. In this context, we can think of data as a raw material consisting of information of value.

This raw data in itself could be extremely useful. Interesting insights can be obtained by processing this data and combining it with other types of data. Data processing helps turn this information into insights that can guide us in making decisions. To illustrate this with a simple example, consider the end-of-day stock price of a publicly traded US company. By observing it for a sufficiently long period, you can see some interesting patterns in stock price changes occurring over time. Similarly, you can observe the performance of the broader S&P 500 index during the same period and compare the performance of the two. Say that an investor was considering whether to invest in stocks of this company or wait until a later time. This decision will involve making use of other types of information, such as an individual's risk tolerance, financial market conditions, and so on.

In the real world, systems that generate data are not perfect. There will be times when the data will have missing elements. Duplicate data is also a common occurrence. In essence, before the actual data analysis work can be started, the raw data must be collected, cleaned, and organized so that it is suitable for further analysis.

The following are the topics that we will be covering in this chapter:

- Data journey
- Sourcing data
- Understanding data
- Using machine learning to learn from data
- Creating a data pipeline

Data journey

Let's look at the journey of data from its creation to its usage:

- **Raw data creation**: The observation, event, action, and manual entry are the key elements that contribute to data creation. This data is typically persisted as a raw data source for future usage. The persistent storage could be flat files, a database, a Kafka topic, AWS **Kinesis Data Streams** (**KDS**), or any other suitable storage.
- **Raw data extraction**: Raw data extraction is the act of receiving or fetching raw data from a source. In an enterprise, raw data sources are internal as well as external. Some examples of commonly used external sources are currency exchange rates, stock prices, and weather data. A company's transactional data is an example of internal data.
- **Raw data ingestion**: Raw data ingestion refers to the act of storing raw data in an organized form to support orderly data extraction and consumption. Often, this step also involves converting data into a machine-friendly form. For example, if the raw source data is a log message, it could be beneficial to extract the relevant pieces of data from it and save it as structured information for further consumption.
- **Data cleaning and validation**: Data cleaning and validation involves sanitizing data to clean up errors and remove noise. This step could also involve validating the integrity of the data received. This step is particularly crucial for big data-like sources, where a small fraction of data errors is expected. Enterprise-wide appropriate data policies concerning data errors and mitigation are necessary to deal with this scenario.
- **Data enrichment**: Data enrichment involves adding related information by looking up values using standardization services—for example, determining the associated value of a ZIP code plus four, given a US street address.
- **Combining datasets**: Datasets can be joined based on their relationship to each other—for example, joining a customer's purchases with their billing and shipping address at the time of purchase.

- **Analyzing**: When we analyze the data, we look for patterns to discover the relationship—for example, looking at the price of crude oil versus economic activity during the same period.
- **Visualization**: During visualization, we use visual tools to understand patterns and relationships—for example, using graphs to compare crude oil prices and economic activity side by side and varying the period of time that is analyzed.
- **Building hypothesis**: Representing the pattern/relationship of the data and the factors and variables involved is crucial in establishing a hypothesis—for example, establishing a hypothesis that an uptick in economic activity results in an increase in the price of crude oil.
- **Testing hypothesis**: Testing the hypothesis means confirming the validity of an established hypothesis—for example, testing the increase in crude oil prices hypothesis against a larger dataset of crude oil prices and economic activity.
- **Pipeline creation**: The creation of a pipeline involves putting the related tasks together to create a pipeline of work. This is a concrete realization of one or more of the steps outlined previously to create an end-to-end solution.

The data journey outlined in the preceding list is not necessarily sequential and linear in nature. There is general feedback and a refinement of the process that keeps on taking place until you are ready for a pipeline creation. In today's data-driven world, **machine learning** (**ML**) plays a significant role in going through some of these steps, such hypothesis building and testing. Next, we will look at an example of data analysis involving a life cycle task, focusing on the following broad categories:

- Sourcing data
- Understanding data
- Using ML to learn from data
- Creating a data pipeline

Sourcing data

Data sourcing is one of the most preliminary steps in the life cycle of data. It includes activities such as data acquisition, cleaning, and organization. The following is a list of the specific activities that it involves:

- Raw data delivery—push model versus pull (extract) model
- Handling a variety of data formats (CSV, JSON, XML)
- Detecting errors in the data that is delivered
- Removing bad data

- Data enrichment—filling the gaps in the data
- Combining data with other datasets
- Defining a data model
- Transforming the raw data model into the defined model
- Storing the data

Data formats

By its very nature, there is a wide variety of raw data. There are a variety of systems generating raw data for a variety of purposes. The format of data also varies in each of these cases. To perform meaningful data analysis, we need to deal with these various formats and variables effectively. In this section, we will begin by looking at the three most prevalent data formats: XML, JSON, and CSV.

XML

Extensible Markup Language (**XML**) is a simple and popular data format that is commonly used for exchanging data between the enterprise system (https://www.w3.org/xml/). Scala has excellent support for processing XML data using the scala-xml library.

Let's start by adding dependencies to the scala-xml library to build.sbt, as shown in the following code:

```
libraryDependencies ++= Seq(
  "org.scala-lang.modules" %% "scala-xml" % "1.1.0" // Scala XML library
)
```

Let's explore some of the features of how this library works, as shown in the following code:

1. Import Elem from scala.xml package:

   ```
   scala> import scala.xml.Elem
   import scala.xml.Elem
   ```

2. Define a Person case class:

   ```
   scala> case class Person(id: String, fname: String, lname: String,
   age: Option[Int] = None) // class for holding Person object
   defined class Person
   ```

3. Create an XML message:

```
scala> val personXml: Elem = <person
id="123"><fname>John</fname><lname>Doe</lname><age>21</age></person
> // sample XML data
personXml: scala.xml.Elem = <person
id="123"><fname>John</fname><lname>Doe</lname><age>21</age></person
>
```

4. Extract the `id` XML attribute from XML message:

```
scala> val id = personXml \@ "id" // extract XML attribute
id: String = 123
```

5. Extract `fname` XML element from XML message:

```
scala> val fname = personXml \ "fname" // XML element extraction
fname: scala.xml.NodeSeq = NodeSeq(<fname>John</fname>)
```

6. Extract `lname` XML element from XML message:

```
scala> val lname = personXml \ "lname"
lname: scala.xml.NodeSeq = NodeSeq(<lname>Doe</lname>)
```

7. Extract `age` XML element from XML message:

```
scala> val age = personXml \ "age"
age: scala.xml.NodeSeq = NodeSeq(<age>21</age>)
```

8. Now construct `Person` object from extracted values:

```
scala> val person = Person(id, fname.text, lname.text,
Some(age.text.toInt)) // to extract value from element, we need to
use text
person: Person = Person(123,John,Doe,Some(21))
```

As you can see, the `scala-xml` library makes it really convenient to parse XML data. Creating XML is equally straightforward, as illustrated in the following code:

```
scala> import scala.xml.Elem
import scala.xml.Elem

scala> case class Person(id: String, fname: Strig, lname: String, age:
Option[Int] = None)
defined class Person

scala> def toXml(p: Person): Elem = { <person
id={p.id}><fname>{p.fname}</fname><lname>{p.lname}</lname><age>{p.age.getOr
```

```
Else(-1)}</age></person> }
toXml: (p: Person)scala.xml.Elem

scala> val person = Person("123", "John", "Doe", Some(21))
person: Person = Person(123,John,Doe,Some(21))

scala> toXml(person)
res0: scala.xml.Elem = <person
id="123"><fname>John</fname><lname>Doe</lname><age>21</age></person>
```

Spark has excellent support for processing XML data files using the `spark-xml` library (`https://github.com/databricks/spark-xml`). We will be covering this in upcoming chapters on Spark and distributed processing. In addition, Java's native and add-on libraries have excellent support for XML processing, which can also be used with Scala code seamlessly.

JSON

JavaScript Object Notation (JSON) `http://json.org/` is an extremely popular data format. One of the greatest advantages of this format is its simplicity and programming-language neutrality.

We will use the Scala `json4s` (`http://json4s.org/`) library to work with JSON data. We will be using a native library that is the same or similar to the Scala `Lift` library.

Let's set up `build.sbt` with the following dependency and restart SBT using the following code:

```
libraryDependencies ++= Seq(
  "org.json4s" %% "json4s-native" % "3.6.1" // Scala Lift JSON Library
)
```

In the Scala REPL console, explore the library using Scala REPL, as shown in the following code:

```
scala> import org.json4s._
import org.json4s._

scala> import org.json4s.native.JsonMethods._
import org.json4s.native.JsonMethods._

scala> implicit val formats = DefaultFormats
formats: org.json4s.DefaultFormats.type =
org.json4s.DefaultFormats$@59db09a7
```

```
scala> case class Person(id: String, fname: String, lname: String, age:
Int)
defined class Person

scala> val personStr = """{
     | "id": "123",
     | "fname": "John",
     | "lname": "Doe",
     | "age": 21
     | }"""
personStr: String =
{
  "id": "123",
  "fname": "John",
  "lname": "Doe",
  "age": 21
}

scala> val json = parse(personStr) // parses JSON string to JValue
json: org.json4s.JValue = JObject(List((id,JString(123)),
(fname,JString(John)), (lname,JString(Doe)), (age,JInt(21))))

scala> val person = json.extract[Person] // convert to Person object
person: Person = Person(123,John,Doe,21)
```

As you can see in the preceding code, parsing JSON is fairly straightforward using this library. Another great feature of the library is that a parsed JSON object can be easily extracted into a Scala case class object.

Creating JSON is equally straightforward, as shown in the following code:

```
scala> import org.json4s.native.JsonMethods._
import org.json4s.native.JsonMethods._

scala> import org.json4s.JsonDSL._
import org.json4s.JsonDSL._

scala> val json = ("id" -> "123") ~ ("fname" -> "John") ~ ("lname" ->
"Doe") ~ ("age" -> 21) // build JSON object
json: org.json4s.JsonAST.JObject = JObject(List((id,JString(123)),
(fname,JString(John)), (lname,JString(Doe)), (age,JInt(21))))

scala> compact(render(json)) // create compact JSON string
res0: String = {"id":"123","fname":"John","lname":"Doe","age":21}
```

Spark has excellent built-in support for processing JSON data files. We will be covering this in upcoming chapters on Spark and distributed processing. There are also several excellent Java libraries supporting JSON processing that can be easily integrated into Scala code.

CSV

Comma-separated values (CSV) is another very popular data format. Although at first the format appears to be very simple, there are a significant number of edge cases where fairly sophisticated parsing is needed to parse CSV data. There are excellent libraries written in Java and many other languages that have been created to parse and generate CSV data.

We will be looking at how to handle CSV parsing using the Apache commons CSV Java library. This example will also demonstrate how easily Java libraries can be used with Scala code.

First, let's set up our `build.sbt` with the appropriate dependency using the following code:

```
libraryDependencies ++= Seq(
  "org.apache.commons" % "commons-csv" % "1.6" // Apache Commons CSV Java
Library
)
```

Rerun SBT and explore the following code in Scala code. In this example, we are using a dataset that is available as part of the US government's Open Data initiative. This dataset is related to the 2010 census data of the City of Los Angeles in California:

```
scala> import java.io.{BufferedReader, InputStreamReader}
import java.io.{BufferedReader, InputStreamReader}

scala> import java.util.function.Consumer
import java.util.function.Consumer

scala> import org.apache.commons.csv.{CSVFormat, CSVRecord}
import org.apache.commons.csv.{CSVFormat, CSVRecord}

scala> import scala.collection.mutable.ListBuffer
import scala.collection.mutable.ListBuffer

scala> case class CensusData(zipCode: String, totalPopulation: Int,
medianAge: Double,
     | totalMales: Int, totalFemales: Int, totalHouseholds: Int,
averageHouseholdSize: Double)
defined class CensusData

scala> class DataConsumer extends Consumer[CSVRecord] {
     | val buf = ListBuffer[CensusData]()
     | override def accept(t: CSVRecord): Unit = {
     | buf += CensusData(t.get(0), t.get(1).toInt, t.get(2).toDouble,
     | t.get(3).toInt, t.get(4).toInt, t.get(5).toInt, t.get(6).toDouble)
     | }
```

```
    | }
defined class DataConsumer

scala> val reader = new BufferedReader(
    | new InputStreamReader(
    | new
java.net.URL("https://data.lacity.org/api/views/nxs9-385f/rows.csv?accessTy
pe=DOWNLOAD").openStream()
    | )
    | )
reader: java.io.BufferedReader = java.io.BufferedReader@4c83902

scala> val csvParser =
CSVFormat.RFC4180.withFirstRecordAsHeader().parse(reader)
csvParser: org.apache.commons.csv.CSVParser =
org.apache.commons.csv.CSVParser@543b331c

scala> val dataConsumer = new DataConsumer
dataConsumer: DataConsumer = DataConsumer@3254eeb8

scala> csvParser.forEach(dataConsumer)

scala> val allRecords = dataConsumer.buf.toList
allRecords: List[CensusData] = List(CensusData(91371,1,73.5,0,1,1,1.0),
CensusData(90001,57110,26.6,28468,28642,12971,4.4),
CensusData(90002,51223,25.5,24876,26347,11731,4.36),
CensusData(90003,66266,26.3,32631,33635,15642,4.22),
CensusData(90004,62180,34.8,31302,30878,22547,2.73),
CensusData(90005,37681,33.9,19299,18382,15044,2.5),
CensusData(90006,59185,32.4,30254,28931,18617,3.13),
CensusData(90007,40920,24.0,20915,20005,11944,3.0),
CensusData(90008,32327,39.7,14477,17850,13841,2.33),
CensusData(90010,3800,37.8,1874,1926,2014,1.87),
CensusData(90011,103892,26.2,52794,51098,22168,4.67),
CensusData(90012,31103,36.3,19493,11610,10327,2.12),
CensusData(90013,11772,44.6,7629,4143,6416,1.26),
CensusData(90014,7005,44.8,4471,2534,4109,1.34), CensusData(90015,18986,...

scala> allRecords.take(3).foreach(println) // Output first 3 records
CensusData(91371,1,73.5,0,1,1,1.0)
CensusData(90001,57110,26.6,28468,28642,12971,4.4)
CensusData(90002,51223,25.5,24876,26347,11731,4.36)
```

As can be seen in the preceding code, we are able to get CSV data from the URL and then parse it into the Scala case class. Generating CSV data is even more straightforward.

Let's go ahead and generate some CSV data using the following code:

```scala
scala> import org.apache.commons.csv.{CSVFormat, CSVPrinter}
import org.apache.commons.csv.{CSVFormat, CSVPrinter}

scala> val csvPrinter = new CSVPrinter(System.out,
CSVFormat.RFC4180.withHeader("fname", "lname", "age"))
fname,lname,age
csvPrinter: org.apache.commons.csv.CSVPrinter =
org.apache.commons.csv.CSVPrinter@7ff05a74

scala> csvPrinter.printRecord("Jon", "Doe", "21")
Jon,Doe,21

scala> csvPrinter.printRecord("James", "Bond", "39")
James,Bond,39

scala> csvPrinter.flush()
```

Spark has excellent built-in support for processing CSV data files. We will be covering this in upcoming chapters on Spark and distributed processing.

Understanding data

Data generally tells a story. However, this is not obvious just from looking at the data. To understand data, we need to be able to ask certain questions and get answers from the data. Asking the right questions in itself requires a great deal of domain knowledge and experience. Once the questions are framed, getting the answers from the data is the next crucial task. Data exploration is an iterative journey because getting answers to questions generally leads to more questions, and then one has to answer these new questions using data.

We will look at the following two important techniques for understanding and exploring data:

- **Statistical methods**: Looking at the properties of data at an aggregate level
- **Visual methods**: Looking at the properties of data using visual methods

In fact, in many real scenarios, both of these methods are used in conjunction with each other to explore data in an effective manner.

Using statistical methods for data exploration

In this section, we will explore data by looking at some aggregate-level information about the dataset. In a large enough dataset, looking at every individual record and trying to get insight could be a fairly time-consuming process. Statistical methods can help to speed up this process because we can leverage machines for fast and efficient computation aggregates.

We will first use pure Scala code to explore and get an insight into the data. Next, we will look at some Scala libraries that simplify this task even further.

Using Scala

Let's explore the same dataset from the US government's Open Data initiative that we used for our CSV example. Let's make sure that the sbt dependency is defined as follows:

```
libraryDependencies ++= Seq(
  "org.apache.commons" % "commons-csv" % "1.6" // Apache Commons CSV
                                                       Java Library
)
```

Launch your SBT and start the Scala console. For the sake of clarity, all of the steps for processing the CSV have been repeated.

Import the required libraries using the following code:

```
scala> import java.io.{BufferedReader, InputStreamReader}
import java.io.{BufferedReader, InputStreamReader}

scala> import java.util.function.Consumer
import java.util.function.Consumer

scala> import org.apache.commons.csv.{CSVFormat, CSVRecord}
import org.apache.commons.csv.{CSVFormat, CSVRecord}

scala> import scala.collection.mutable.ListBuffer
import scala.collection.mutable.ListBuffer
```

Let's move ahead and write our main code, as follows:

```
scala> case class CensusData(zipCode: String, totalPopulation: Int,
medianAge: Double,
     | totalMales: Int, totalFemales: Int, totalHouseholds: Int,
averageHouseholdSize: Double)
defined class CensusData
```

```
scala> class DataConsumer extends Consumer[CSVRecord] {
     | val buf = ListBuffer[CensusData]()
     | override def accept(t: CSVRecord): Unit = {
     | buf += CensusData(t.get(0), t.get(1).toInt, t.get(2).toDouble,
     | t.get(3).toInt, t.get(4).toInt, t.get(5).toInt, t.get(6).toDouble)
     | }
     | }
defined class DataConsumer

scala> val reader = new BufferedReader(
     | new InputStreamReader(
     | new
java.net.URL("https://data.lacity.org/api/views/nxs9-385f/rows.csv?accessTy
pe=DOWNLOAD").openStream()
     | )
     | )
reader: java.io.BufferedReader = java.io.BufferedReader@572caa8b

scala> val csvParser =
CSVFormat.RFC4180.withFirstRecordAsHeader().parse(reader)
csvParser: org.apache.commons.csv.CSVParser =
org.apache.commons.csv.CSVParser@19405f70

scala> val dataConsumer = new DataConsumer
dataConsumer: DataConsumer = DataConsumer@20d9ee6f

scala> csvParser.forEach(dataConsumer)

scala> val allRecords = dataConsumer.buf.toList
allRecords: List[CensusData] = List(CensusData(91371,1,73.5,0,1,1,1.0),
CensusData(90001,57110,26.6,28468,28642,12971,4.4),
CensusData(90002,51223,25.5,24876,26347,11731,4.36),
CensusData(90003,66266,26.3,32631,33635,15642,4.22),
CensusData(90004,62180,34.8,31302,30878,22547,2.73),
CensusData(90005,37681,33.9,19299,18382,15044,2.5),
CensusData(90006,59185,32.4,30254,28931,18617,3.13),
CensusData(90007,40920,24.0,20915,20005,11944,3.0),
CensusData(90008,32327,39.7,14477,17850,13841,2.33),
CensusData(90010,3800,37.8,1874,1926,2014,1.87),
CensusData(90011,103892,26.2,52794,51098,22168,4.67),
CensusData(90012,31103,36.3,19493,11610,10327,2.12),
CensusData(90013,11772,44.6,7629,4143,6416,1.26),
CensusData(90014,7005,44.8,4471,2534,4109,1.34), CensusData(90015,18986,...
```

Record the analysis using the following code:

```
scala> // Records Analysis

scala> allRecords.size // total records
res1: Int = 319

scala> allRecords.distinct.size // distinct records
res2: Int = 319

scala> allRecords.take(3) // 3 records from dataset
res3: List[CensusData] = List(CensusData(91371,1,73.5,0,1,1,1.0),
CensusData(90001,57110,26.6,28468,28642,12971,4.4),
CensusData(90002,51223,25.5,24876,26347,11731,4.36))

scala> // Zip Code Analysis

scala> allRecords.map(_.zipCode).distinct.size // distinct zipCode
res4: Int = 319

scala> allRecords.map(_.zipCode).min // minimum zipCode
res5: String = 90001

scala> allRecords.map(_.zipCode).max
res6: String = 93591

scala> val averageZip = allRecords.map(_.zipCode).aggregate(0)((a, b) => a
+ b.toInt, (x, y) => x + y) / allRecords.size
averageZip: Int = 91000

scala> allRecords.map(_.zipCode.toInt).sum /allRecords.size // another way
to compute the same
res7: Int = 91000
```

Perform the total population analysis using the following code:

```
scala> // Total Population Analysis

scala> allRecords.map(_.totalPopulation).sum
res8: Int = 10603988

scala> val averagePop = allRecords.map(_.totalPopulation).sum /
allRecords.size
averagePop: Int = 33241

scala> allRecords.sortBy(_.totalPopulation).head // record with lowest
Population
res9: CensusData = CensusData(90079,0,0.0,0,0,0,0.0)
```

```
scala> allRecords.sortBy(-_.totalPopulation).head // record with highest
Population
res10: CensusData = CensusData(90650,105549,32.5,52364,53185,27130,3.83)

scala> // Aggregate total numbers using a single aggregate method

scala> val (totalPopulation, totalMales, totalFemales, totalHouseholds) =
allRecords.aggregate((0, 0, 0, 0))((a, b) => (a._1 + b.totalPopulation,
a._2 + b.totalMales, a._3 + b.totalFemales, a._4 + b.totalHouseholds),
(x,y) => (x._1 + y._1, x._2 + y._2, x._3 + y._3, x._4 + y._4))
totalPopulation: Int = 10603988
totalMales: Int = 5228909
totalFemales: Int = 5375079
totalHouseholds: Int = 3497698
```

As can be seen in the preceding code, the Scala collection API comes in handy when performing data analysis. Also, note the `aggregate` method of the API; it is a generalized way to create an aggregated value over a collection. Let's look at some more ways to create aggregate values in Scala, as shown in the following code:

```
scala> // Aggregate using foldLeft

scala> allRecords.map(_.totalPopulation).foldLeft(0)(_+_)
res11: Int = 10603988

scala> allRecords.map(_.totalMales).foldLeft(0)(_+_)
res12: Int = 5228909

scala> allRecords.map(_.totalFemales).foldLeft(0)(_+_)
res13: Int = 5375079

scala> allRecords.map(_.totalHouseholds).foldLeft(0)(_+_)
res14: Int = 3497698

scala> // Aggregate using foldRight

scala> allRecords.map(_.totalPopulation).foldRight(0)(_+_)
res15: Int = 10603988

scala> allRecords.map(_.totalMales).foldRight(0)(_+_)
res16: Int = 5228909

scala> allRecords.map(_.totalFemales).foldRight(0)(_+_)
res17: Int = 5375079

scala> allRecords.map(_.totalHouseholds).foldRight(0)(_+_)
res18: Int = 3497698
```

```
scala> // Aggregate using reduce

scala> allRecords.map(_.totalPopulation).reduce(_+_)
res19: Int = 10603988

scala> allRecords.map(_.totalMales).reduce(_+_)
res20: Int = 5228909

scala> allRecords.map(_.totalFemales).reduce(_+_)
res21: Int = 5375079

scala> allRecords.map(_.totalHouseholds).reduce(_+_)
res22: Int = 3497698
```

 Note that we are getting the same results using the `foldLeft`, `foldRight`, and `reduce` methods.

Other Scala tools

Spark is a very popular distributed data-processing engine. It has built-in support for exploring data in many different formats. We will look at Spark functionality in subsequent chapters. Let's look at another Scala library called **Saddle** (`http://saddle.github.io/`) and see how we can leverage this library to work with data.

This library is not yet available for Scala 2.12, so we will be using Scala 2.11.12 to explore this library. Configure your `sbt` `build.sbt` file as follows:

```
scalaVersion := "2.11.12"

libraryDependencies ++= Seq(
  "org.scala-saddle" %% "saddle-core" % "1.3.4"
)
```

For this exploration using Saddle, we will continue to use the dataset that we used in our earlier exercise. In your `sbt` console, try the following:

```
scala> import java.io.{BufferedReader, InputStreamReader}
import java.io.{BufferedReader, InputStreamReader}

scala> import org.saddle.io._
import org.saddle.io._

scala> class SaddleCsvSource(url: String) extends CsvSource {
     |   val reader = new BufferedReader(new InputStreamReader(new
```

```
java.net.URL(url).openStream())) 
    | override def readLine: String = { 
    | reader.readLine() 
    | } 
    | } 
defined class SaddleCsvSource

scala> val file = new 
SaddleCsvSource("https://data.lacity.org/api/views/nxs9-385f/rows.csv?acces
sType=DOWNLOAD")
file: SaddleCsvSource = SaddleCsvSource@6437b766

scala> val frame = CsvParser.parse(file)
frame: org.saddle.Frame[Int,Int,String] =
[320 x 7]
              0 1 2 3 4 5 6
-------- ---------------- ---------- ----------- -------------- --------
  0 -> Zip Code Total Population Median Age Total Males Total Females Total
Households Average Household Size
  1 -> 91371 1 73.5 0 1 1 1
  2 -> 90001 57110 26.6 28468 28642 12971 4.4
  3 -> 90002 51223 25.5 24876 26347 11731 4.36
  4 -> 90003 66266 26.3 32631 33635 1564...
scala> frame.print() // prints 10 records from the frame
[320 x 7]
              0 1 2 3 4 5 6
-------- ---------------- ---------- ----------- -------------- --------
  0 -> Zip Code Total Population Median Age Total Males Total Females Total
Households Average Household Size
  1 -> 91371 1 73.5 0 1 1 1
  2 -> 90001 57110 26.6 28468 28642 12971 4.4
  3 -> 90002 51223 25.5 24876 26347 11731 4.36
  4 -> 90003 66266 26.3 32631 33635 15642 4.22
...
315 -> 93552 38158 28.4 18711 19447 9690 3.93
316 -> 93553 2138 43.3 1121 1017 816 2.62
317 -> 93560 18910 32.4 9491 9419 6469 2.92
318 -> 93563 388 44.5 263 125 103 2.53
319 -> 93591 7285 30.9 3653 3632 1982 3.67
```

If you are familiar with R or Python's pandas library, you will find a great deal of similarity between Saddle's API and these APIs. The `frame` object that we constructed previously lets us work with the data at a higher level of abstraction using Saddle's API. Let's further explore Saddle's frame API, as shown in the following code:

```
scala> val df = frame.withColIndex(0) // first row is the CSV header
df: org.saddle.Frame[Int,String,String] = [319 x 7]
      Zip Code Total Population Median Age Total Males Total Females Total
```

```
Households Average Household Size
-------- ------------------ ---------- ----------- -------------- --------
    1 -> 91371 1 73.5 0 1 1 1
    2 -> 90001 57110 26.6 28468 28642 12971 4.4
    3 -> 90002 51223 25.5 24876 26347 11731 4.36
    4 -> 90003 66266 26.3 32631 33635 15642 4.22
    5 -> 90004 62180 34.8 31302 30878 2254...
scala> df.col("Zip Code") // we can access each column by name
res1: org.saddle.Frame[Int,String,String] =
[319 x 1]
        Zip Code
        --------
    1 -> 91371
    2 -> 90001
    3 -> 90002
    4 -> 90003
    5 -> 90004
...
  315 -> 93552
  316 -> 93553
  317 -> 93560
  318 -> 93563
  319 -> 93591

scala> df.col("Zip Code").min // should fail
<console>:17: error: No implicit view available from org.saddle.Series[_,
String] => org.saddle.stats.VecStats[String].
        df.col("Zip Code").min // should fail
                          ^

scala> df.col("Zip Code").mapValues(CsvParser.parseInt).min // convert from
string to integer
res3: org.saddle.Series[String,Int] =
[1 x 1]
Zip Code -> 90001
```

The preceding example demonstrates how to work with a frame that consists of rows and columns. It also shows you how to extract a specific column using the qualified name and compute some simple stats, such as `min`.

Next, let's look at how to get the same information using Saddle. Here, we'll try obtaining the ZIP codes using the following code:

```
scala> df.col("Zip Code").mapValues(CsvParser.parseInt).min
res4: org.saddle.Series[String,Int] =
[1 x 1]
Zip Code -> 90001
```

```
scala> df.col("Zip Code").mapValues(CsvParser.parseInt).max
res5: org.saddle.Series[String,Int] =
[1 x 1]
Zip Code -> 93591

scala> df.col("Zip Code").mapValues(CsvParser.parseInt).mean
res6: org.saddle.Series[String,Double] =
[1 x 1]
Zip Code -> 91000.6740
```

Next, let's obtain the total population using the following code:

```
scala> df.col("Total Population").mapValues(CsvParser.parseInt).min
res7: org.saddle.Series[String,Int] =
[1 x 1]
Total Population -> 0

scala> df.col("Total Population").mapValues(CsvParser.parseInt).max
res8: org.saddle.Series[String,Int] =
[1 x 1]
Total Population -> 105549

scala> df.col("Total Population").mapValues(CsvParser.parseInt).sum
res9: org.saddle.Series[String,Int] =
[1 x 1]
Total Population -> 10603988
```

Next, let's find the total number of males using the following code:

```
scala> df.col("Total Males").mapValues(CsvParser.parseInt).min
res10: org.saddle.Series[String,Int] =
[1 x 1]
Total Males -> 0

scala> df.col("Total Males").mapValues(CsvParser.parseInt).max
res11: org.saddle.Series[String,Int] =
[1 x 1]
Total Males -> 52794

scala> df.col("Total Males").mapValues(CsvParser.parseInt).sum
res12: org.saddle.Series[String,Int] =
[1 x 1]
Total Males -> 5228909
```

Next, let's find the total number of females using the following code:

```
scala> df.col("Total Females").mapValues(CsvParser.parseInt).min
res13: org.saddle.Series[String,Int] =
[1 x 1]
```

```
Total Females -> 0

scala> df.col("Total Females").mapValues(CsvParser.parseInt).max
res14: org.saddle.Series[String,Int] =
[1 x 1]
Total Females -> 53185

scala> df.col("Total Females").mapValues(CsvParser.parseInt).sum
res15: org.saddle.Series[String,Int] =
[1 x 1]
Total Females -> 5375079
```

Now let's find the total number of households using the following code:

```
scala> df.col("Total Households").mapValues(CsvParser.parseInt).min
res16: org.saddle.Series[String,Int] =
[1 x 1]
Total Households -> 0

scala> df.col("Total Households").mapValues(CsvParser.parseInt).max
res17: org.saddle.Series[String,Int] =
[1 x 1]
Total Households -> 31087

scala> df.col("Total Households").mapValues(CsvParser.parseInt).sum
res18: org.saddle.Series[String,Int] =
[1 x 1]
Total Households -> 3497698
```

Saddle's Scala library has a lot more to offer in terms of computing useful statistical information and working with data. Let's implement some other methods supported by Saddle using the following code:

```
scala> df.numRows
res19: Int = 319

scala> df.numCols
res20: Int = 7

scala> df.col("Total Households").mapValues(CsvParser.parseInt).mean
res21: org.saddle.Series[String,Double] =
[1 x 1]
Total Households -> 10964.5705

scala> df.col("Total Households").mapValues(CsvParser.parseInt).median
res22: org.saddle.Series[String,Double] =
[1 x 1]
Total Households -> 10968.0000
```

```
scala> df.col("Total Households").mapValues(CsvParser.parseInt).stdev
res23: org.saddle.Series[String,Double] =
[1 x 1]
Total Households -> 6270.6464
```

Let's see a list of the total number of households, as shown in the following code:

```
// convert to Scala List
scala> df.col("Total
       Households").mapValues(CsvParser.parseInt).toSeq.map(_._3).toList

res24: List[Int] = List(1, 12971, 11731, 15642, 22547, 15044, 18617, 11944,
13841, 2014, 22168, 10327, 6416, 4109, 7420, 16145, 9338, 15493, 23344,
16514, 1561, 17023, 10727, 17903, 21228, 24956, 21929, 14964, 13883, 11156,
12765, 12924, 25592, 12814, 18646, 15869, 11928, 11436, 3317, 9513, 19892,
16075, 25144, 15224, 28534, 16168, 11821, 16657, 3371, 15658, 892, 9596,
6892, 9155, 13260, 10968, 14476, 23985, 1510, 12326, 13364, 0, 4, 3615, 0,
31, 0, 2949, 2, 24104, 8669, 3706, 5567, 12741, 11630, 7520, 12883, 6605,
7632, 13617, 12687, 7085, 15830, 3427, 8880, 31087, 9550, 18419, 10429,
14669, 0, 7174, 14038, 6554, 9212, 9479, 15618, 16910, 16009, 23278, 2612,
14261, 12654, 6575, 11895, 10684, 7290, 6634, 5933, 4188, 5301, 13970,
10089, 14376, 14610, 5717, 17183, 11580, 14244, 0, 11027, ...
```

As we can see from the preceding examples, Saddle's API has a lot to offer in terms of conveniently exploring and working with data.

 Please note that while Saddle works well in a single **Java Virtual Machine (JVM)**, it is not designed for data processing in a distributed environment.

Spark is better for working with data in a distributed environment and for processing data at a large scale. We will look at Spark in subsequent chapters.

Using data visualization for data exploration

Using the data visualization methodology, we can get an insight into the data by looking at the visual representation of the data. We will be looking at a fairly popular Scala library to do some simple exploration in Scala.

Using the vegas-viz library for data visualization

We will explore some sample dates using the `vegas-viz` (https://www.vegas-viz.org/) Scala library for data visualization. This is a powerful Scala library that integrates very well with Spark. We will work with Spark in subsequent chapters.

To explore this library in `sbt`, we will first set up the `build.sbt` file using the following code. At the time of writing, `vegas-viz` and Spark are only supported for Scala 2.11.x, so we will use Scala version 2.11.12 for our exploration:

```
// We will use Scala 2.11.x because many of Scala libraries such as
// Spark, vegas-viz are not yet supported for Scala 2.12.x
scalaVersion := "2.11.12"

libraryDependencies ++= Seq(
  "org.vegas-viz" %% "vegas" % "0.3.11" // Vegas Visualization Library
)
```

After creating the aforementioned `build.sbt`, run SBT. Once inside SBT, run the following console command to start Scala REPL:

```
scala> val plot = Vegas("Currency Exchange Rates").
     | withData(
     | Seq(
     | Map("Currency Code" -> "USD", "Exchange Rate" -> 1.00),
     | Map("Currency Code" -> "EUR", "Exchange Rate" -> 0.86),
     | Map("Currency Code" -> "GBP", "Exchange Rate" -> 0.76),
     | Map("Currency Code" -> "CHF", "Exchange Rate" -> 0.99),
     | Map("Currency Code" -> "CAD", "Exchange Rate" -> 1.29),
     | Map("Currency Code" -> "AUD", "Exchange Rate" -> 1.41),
     | Map("Currency Code" -> "HKD", "Exchange Rate" -> 7.83)
     | )
     | ).
     | encodeX("Currency Code", Nom).
     | encodeY("Exchange Rate", Quant).
     | mark(Point)
```

Done. Now, let's plot the following:

```
plot: vegas.DSL.ExtendedUnitSpecBuilder =
ExtendedUnitSpecBuilder(ExtendedUnitSpec(None,None,Point,Some(Encoding(None
,None,Some(PositionChannelDef(None,None,None,Some(Currency
Code),Some(Nominal),None,None,None,None)),Some(PositionChannelDef(None
,None,None,Some(Exchange
Rate),Some(Quantitative),None,None,None,None)),None,None,None,None,Non
e,None,None,None,None,None)),None,Some(Currency Exchange
Rates),Some(Data(None,None,Some(List(Values(Map(Currency Code -> USD,
```

```
Exchange Rate -> 1.0)), Values(Map(Currency Code -> EUR, Exchange Rate ->
0.86)), Values(Map(Currency Code -> GBP, Exchange Rate -> 0.76)),
Values(Map(Currency Code -> CHF, Exchange Rate -> 0.99)),
Values(Map(Currency Code -> CAD, Exchange Rate -> 1.29)),
Values(Map(Currency Code -> AUD, Exchange Rate -> 1.41)), ...
scala> plot.show
```

This will produce the following scatter plot of **Currency Code** versus **Exchange Rate** (USD):

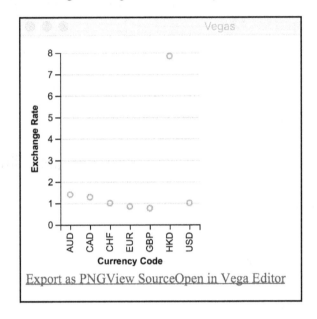

Change the plot mark to `Bar` to output the bar chart using the following code:

```
val plot = Vegas("Currency Exchange Rates").
  ...
  mark(Bar) // for bar chart
```

This produces the following bar chart:

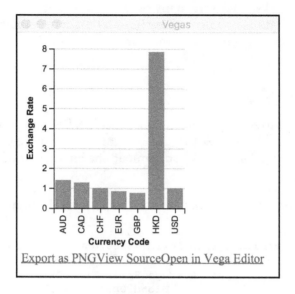

As we can see, by using the `vegas-viz` Scala library, we can easily perform data visualization using a simple set of APIs.

Other libraries for data visualization

There are several other libraries that support data visualization, but what makes the `vegas-viz` library an important player is that it has excellent integration with Spark. Spark has no built-in support for data visualization, and the `vegas-viz` library fills in this gap very nicely.

One tool that you should definitely look at is Jupyter Notebook (`http://jupyter.org/`), which aims to provide an integrated environment for performing sophisticated data analysis, and has excellent support for visual libraries.

Using ML to learn from data

ML is the process of discovering patterns in data without explicitly programming the logic. In this section, we will look at how ML can be performed using Scala. This is a very vast topic in itself, so we will only look at this from an applied usage point of view.

Smile (`https://haifengl.github.io/smile/index.html`) is a popular Scala library that helps with ML-related tasks. There are many other Scala ML libraries that are extremely popular, such as Spark MLlib; however, these libraries are more suitable for distributed processing environments. We will look at these in subsequent chapters.

Setting up Smile

There are multiple ways to set up Smile on your machine. Refer to Smile Quick Start at `https://haifengl.github.io/smile/quickstart.html` for more information. The easiest and fastest way to get started is to download the binaries from `https://github.com/haifengl/smile/releases`.

The following is a set of commands that are used to perform the Smile setup:

```
$ curl -L -o smile-1.5.1.zip
https://github.com/haifengl/smile/releases/download/v1.5.1/smile-1.5.1.zip
# download and save in zip file
  % Total % Received % Xferd Average Speed Time Time Time Current
                               Dload Upload Total Spent Left Speed
100 605 0 605 0 0 1729 0 --:--:-- --:--:-- --:--:-- 1733
100 165M 100 165M 0 0 11.2M 0 0:00:14 0:00:14 --:--:-- 12.2M

$ unzip smile-1.5.1.zip # unzip file
Archive: smile-1.5.1.zip
  inflating: smile-1.5.1/smile_config.txt
  inflating: smile-1.5.1/bin/init.scala
  inflating: smile-1.5.1/bin/libblas3.dll
. . .
  inflating: smile-1.5.1/lib/com.github.javaparser.javaparser-
core-3.2.5.jar
  inflating: smile-1.5.1/lib/com.github.scopt.scopt_2.12-3.5.0.jar
  inflating: smile-1.5.1/bin/smile
  inflating: smile-1.5.1/bin/smile.bat

$ cd smile-1.5.1 # Smile is setup in this directory

$ ls -1 # List the contents
bin
conf
data
doc
examples
lib
smile_config.txt
```

```
$ ls -l bin/smile # This is the Smile start up script
-rwxr-xr-x 1 uid gid 12980 Feb 25 2018 bin/smile
```

Once the setup is complete, let's confirm that it is working properly using the following code. Note that there is a JVM memory parameter that might have to be adjusted depending upon the size of the dataset that is being worked on:

```
$ ./bin/smile -J-Xmx2048M # 2048M (2 GB) of memory to JVM
Compiling (synthetic)/ammonite/predef/interpBridge.sc
Compiling (synthetic)/ammonite/predef/replBridge.sc
Compiling (synthetic)/ammonite/predef/DefaultPredef.sc
Compiling (synthetic)/ammonite/predef/CodePredef.sc
```

```
                                                   ..::'''':::..
                                               .;'''   ``;.

         .... :: :: :: ::
       ,;' .;: () ..: :: :: :: ::
      ::. ..:,:;.,:;. . :: .:::. :: .:' :: :: `:. ::
       '''::, :: :: :: `:: :: ;: .:: :: : : ::
     ,:'; :::; :: :: :: :: :: ::,::''. :: `:. .:' ::
     `:,,,,,ii' ,ii ,iii ii, ,iii ,ii, `:,,,,:' `;..``::::''..;'
                                                   ``::,,,,::''

       Welcome to Smile Shell; enter 'help<RETURN>' for the list of commands.
       Type "exit<RETURN>" to leave the Smile Shell
       Version 1.5.1, Scala 2.12.4, SBT 1.1.0, Built at 2018-02-26
                                  02:31:25.456
=============================================================================
```

Let's see what things can be done using the Smile shell, as shown in the following code:

```
smile> help

  General:
    help -- print this summary
    :help -- print Scala shell command summary
    :quit -- exit the shell
    demo -- show demo window
    benchmark -- benchmark tests

  I/O:
    read -- Reads an object/model back from a file created by write command.
  ...
  Classification:
    knn -- K-nearest neighbor classifier.
    logit -- Logistic regression.
  ...
  Regression:
```

```
   ols -- Ordinary least square.
   ridge -- Ridge regression.
   lasso -- Least absolute shrinkage and selection operator.
 ...
 Graphics:
   plot -- Scatter plot.
   line -- Scatter plot which connects points by straight lines.
   boxplot -- Boxplots can be useful to display differences between
populations.
 ...
```

As can be seen from the `help` message, Smile supports a wide range of classification and regression ML algorithms. Another nice feature of Smile is that it also has support for data visualization.

Running Smile

To explore Smile, we will run some of the examples that are included with the Smile code base.

The following is an example of applying a random forest algorithm to the data:

```
smile> val data = read.arff("data/weka/iris.arff", 4)
data: AttributeDataset = iris
   class sepallength sepalwidth petallength petalwidth
[1] Iris-setosa 5.1000 3.5000 1.4000 0.2000
[2] Iris-setosa 4.9000 3.0000 1.4000 0.2000
[3] Iris-setosa 4.7000 3.2000 1.3000 0.2000
[4] Iris-setosa 4.6000 3.1000 1.5000 0.2000
[5] Iris-setosa 5.0000 3.6000 1.4000 0.2000
[6] Iris-setosa 5.4000 3.9000 1.7000 0.4000
[7] Iris-setosa 4.6000 3.4000 1.4000 0.3000
[8] Iris-setosa 5.0000 3.4000 1.5000 0.2000
[9] Iris-setosa 4.4000 2.9000 1.4000 0.2000
[10] Iris-setosa 4.9000 3.1000 1.5000 0.1000
140 more rows...

smile> val (x, y) = data.unzipInt
x: Array[Array[Double]] = Array(
   Array(5.1, 3.5, 1.4, 0.2),
   Array(4.9, 3.0, 1.4, 0.2),
   Array(4.7, 3.2, 1.3, 0.2),
   Array(4.6, 3.1, 1.5, 0.2),
 ...
smile> val rf = randomForest(x, y)
[Thread-209] INFO smile.classification.RandomForest - Random forest tree
```

```
OOB size: 59, accuracy: 89.83%
[Thread-210] INFO smile.classification.RandomForest - Random forest tree
OOB size: 52, accuracy: 88.46%
[Thread-213] INFO smile.classification.RandomForest - Random forest tree
OOB size: 50, accuracy: 100.00%
...
[Thread-210] INFO smile.classification.RandomForest - Random forest tree
OOB size: 56, accuracy: 100.00%
[main] INFO smile.util.package$ - runtime: 97.07988 ms
rf: RandomForest = smile.classification.RandomForest@a4df251

smile> println(s"OOB error = ${rf.error}")
OOB error = 0.04666666666666667


smile> rf.predict(x(0))
res4: Int = 0
```

Let's explore the data using Smile's visualization features, as shown in the following code:

```
smile> val iris = read.arff("data/weka/iris.arff", 4)
iris: AttributeDataset = iris
  class sepallength sepalwidth petallength petalwidth
[1]  Iris-setosa 5.1000 3.5000 1.4000 0.2000
[2]  Iris-setosa 4.9000 3.0000 1.4000 0.2000
[3]  Iris-setosa 4.7000 3.2000 1.3000 0.2000
[4]  Iris-setosa 4.6000 3.1000 1.5000 0.2000
[5]  Iris-setosa 5.0000 3.6000 1.4000 0.2000
[6]  Iris-setosa 5.4000 3.9000 1.7000 0.4000
[7]  Iris-setosa 4.6000 3.4000 1.4000 0.3000
[8]  Iris-setosa 5.0000 3.4000 1.5000 0.2000
[9]  Iris-setosa 4.4000 2.9000 1.4000 0.2000
[10] Iris-setosa 4.9000 3.1000 1.5000 0.1000
140 more rows...

smile> plot(iris, '*', Array(Color.RED, Color.BLUE, Color.CYAN)) // plot
all the attribute pairs
res1: javax.swing.JFrame =
javax.swing.JFrame[frame0,780,191,1000x1000,invalid,layout=java.awt.BorderL
ayout,title=iris,resizable,normal,defaultCloseOperation=DISPOSE_ON_CLOSE,ro
otPane=javax.swing.JRootPane[,0,22,1000x978,invalid,layout=javax.swing.JRoo
tPane$RootLayout,alignmentX=0.0,alignmentY=0.0,border=,flags=16777673,maxim
umSize=,minimumSize=,preferredSize=],rootPaneCheckingEnabled=true]
```

We will then see the following output window:

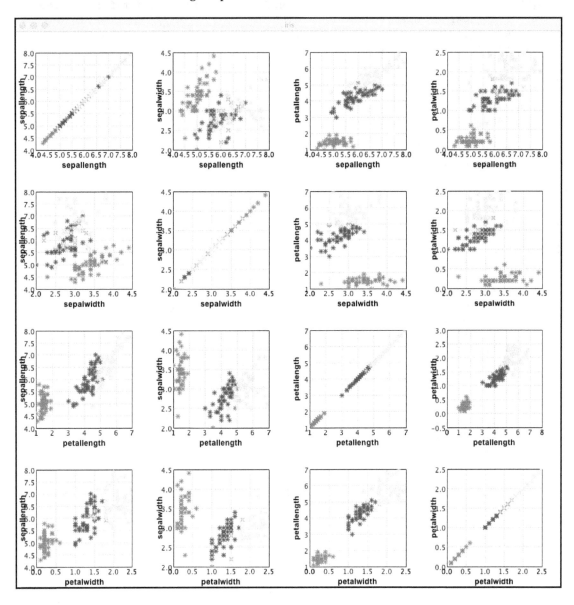

As we can see, the Smile Scala library has a lot to offer in terms of ML data visualization. This could be a tool of choice if the data volumes are not very large. As mentioned earlier, Spark and Vegas would be better for processing and visualizing large datasets.

Creating a data pipeline

We have so far looked at data analysis life cycle tasks in isolation. In the real world, these tasks need to be connected together to create a cohesive solution. Data pipelines are about creating end-to-end, data-oriented solutions.

Spark supports **ML pipelines** (`https://spark.apache.org/docs/2.3.0/ml-pipeline.html`). We will look at Spark and how to use Spark's ML pipeline functionality in subsequent chapters.

Jupyter Notebooks (`http://jupyter.org/`) is another great option for creating an integrated data pipeline. Papermill (`https://github.com/nteract/papermill`) is an open source project that helps parameterize and run Jupyter Notebooks. We will explore some of these options in subsequent chapters.

Summary

In this chapter, we looked at the journey of data and the data analysis life cycle at a broad level. Using hands-on examples, we looked at how to perform some of the tasks using mainly Scala and some Java libraries.

In the next chapter, we will look at data ingestion and associated tasks.

3
Data Ingestion

In this chapter, we will look into the data ingestion aspects of the data life cycle. Data ingestion is a very broad term; however, we will concentrate on some of the most important aspects of it. Data is generally considered to be ingested when it is ready for usage in enterprise systems. In reality, a significant amount of effort is required to perform data ingestion effectively and efficiently.

Data ingestion typically involves the following key tasks, which are also the topics that will be covered in this chapter:

- Data extraction
- Data staging
- Data cleaning
- Data normalization
- Data enrichment
- Data organization and storage

At times, there are more tasks involved as part of data ingestion. There are also situations where some of these tasks might not be necessary and two or more could be combined into a single task.

Data extraction

Data extraction or delivery is the act of making raw data available to an enterprise system for usage. This raw data could be originating from a system outside of the enterprise or could have been created by an internal system. There are two ways in which data can be delivered:

- **Pull**: Data is fetched by the data consumer from the source system
- **Push**: The data producer delivers the data to the consumer

Both of these mechanisms are used extensively in enterprises for exchanging data. The majority of legacy systems and applications use a pull-based approach for data delivery; however, with the growing need of near real-time availability of data, there is a shift toward push-based data delivery.

Let's look at both of these mechanisms in more detail.

Pull-oriented data extraction

In pull oriented data extraction, the consumer fetches the data from the producer. Some of the commonly used pull mechanisms are as follows:

- **FTP/SFTP**: The producer makes the data available on a **File Transfer Protocol (FTP)** server, and the consumer fetches this data from the FTP server using the FTP protocol. The **Secure File Transfer Protocol (SFTP)** variant uses **Secure Sockets Layer (SSL)/Transport Layer Security (TLS)** to encrypt data in transit.
- **JDBC/ODBC**: The producer makes the data available on a **Relational Database Management System (RDMS)**-like system and the consumer queries this data using JDBC/ODBC interface. The **Java Database Connectivity (JDBC)** interface is generally used by JVM-based consumer applications, while **Open Database Connectivity (ODBC)** is used by most others.

> Please note that there are interfaces other than JDBC/ODBC available for querying legacy database systems.

- **HTTP/HTTPS**: The producer makes the data available using HTTP methods such as GET/POST. The HTTPS variant uses SSL/TLS to encrypt data in transit.
- **Web services**: The producer supports web service APIs to provide access to data.

This diagram illustrates how a **Data Consumer** pulls data from the **Data Producer**:

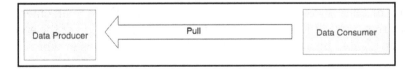

Let's look at a simple and concrete example of pull-based data extraction:

1. Import the `BufferedReader` class from Java's `java.io` package. This class provides an efficient mechanism for reading text data by buffering the data. The details of the `BufferedReader` class can be found at `https://docs.oracle.com/javase/8/docs/api/java/io/BufferedReader.html`:

   ```scala
   scala> import java.io.BufferedReader
   import java.io.BufferedReader
   ```

2. Import the `InputStreamReader` class from Java's `java.io` package. This class provides a bridge from the byte stream to the character stream. The details of the `InputStreamReader` class can be found at `https://docs.oracle.com/javase/8/docs/api/?java/io/InputStreamReader.html`:

   ```scala
   scala> import java.io.InputStreamReader
   import java.io.InputStreamReader
   ```

3. Import URL class from Java's `java.net` package. The URL class represents a **uniform resource locator** reference to a resource on the internet. More details about the URL class can be found at `https://docs.oracle.com/javase/8/docs/api/java/net/URL.html`:

   ```scala
   scala> import java.net.URL
   import java.net.URL
   ```

4. Import the `Consumer` interface from Java's `java.util.function` package. This interface represents an operation that accepts a single input argument and does not return a result. The details of the `Consumer` interface can be found at `https://docs.oracle.com/javase/8/docs/api/java/util/function/Consumer.html`:

   ```scala
   scala> import java.util.function.Consumer
   import java.util.function.Consumer
   ```

5. Import `ListBuffer` from Scala's `scala.collection.mutable` package. `ListBuffer` is a buffered implementation backed by a list. It is to be noted that Scala's List is an immutable structure that, once created, cannot be changed. The `ListBuffer` class provides a convenient mechanism to append items to a buffer and turn it into a `List` object once the appends are completed. The specifications of Scala's `ListBuffer` can be found at `https://www.scala-lang.org/api/current/scala/collection/mutable/ListBuffer.html`:

   ```scala
   scala> import scala.collection.mutable.ListBuffer
   import scala.collection.mutable.ListBuffer
   ```

6. Define a Scala class called `DataConsumer`:
   - This class extends `Consumer`, accepting `String` as input.
   - The `class` constructor initializes a `ListBuffer` of `String` and overrides the `accept` method from `Consumer`.
   - The `accept` method appends the incoming string to the buffer:

```
scala> class DataConsumer extends Consumer[String] {
     | val buf = ListBuffer[String]() // initialize list
                                                   buffer
     | override def accept(t: String): Unit = { // override
                                   //accept method of Consumer
     | buf += t // appends incoming string to list buffer
     | }
     | }
defined class DataConsumer
```

7. Create a new instance of `BufferedReader`, that reads from an `InputStreamReader` pulling data from https://data.lacity.org/api/views/ nxs9-385f/rows.csv?accessType=DOWNLOAD. This involves multiple steps in the following order:
   - First, we create an instance of `java.net.URL` to represent the input network resource. The details of the `java.net.URL` class can be found at https://docs.oracle.com/javase/8/docs/api/java/net/URL. html.
   - Next, invoke the `openStream` method on URL instance to get it as an input byte stream. The `openStream` method returns a Java `java.io.InputStream` instance. This stream represents an input stream of bytes. The details of Java's `InputStream` class can be found at https://docs.oracle.com/javase/8/docs/api/java/io/ InputStream.html.
   - Next, instantiate `InputStreamReader` by wrapping the input byte stream to convert it to a character stream.
   - Finally, instantiate `BufferedReader` by wrapping `InputStreamReader`. This gives us a reader object that can be used for efficiently reading text data line-by-line.

The pattern being used here for instantiating the reader is known as a **decorator design pattern**. Each class is adding additional functionality to the class it is wrapping:

- `InputStreamReader` turns a byte-oriented `InputStream` to a character-oriented input stream.
- `BufferedReader` turns a character-oriented input stream to an input stream that handles data buffering and provides an efficient mechanism with which to read text data.

The following is a complete code for instantiating the `BufferedReader`:

```scala
scala> val reader = new BufferedReader(
     | new InputStreamReader(
     | new URL("https://data.lacity.org/api/views/nxs9-
         385f/rows.csv?accessType=DOWNLOAD").openStream()
     | )
     | )
reader: java.io.BufferedReader = java.io.BufferedReader@558ab02f
```

8. Create a new instance of `DataConsumer`:

```scala
scala> val dataConsumer = new DataConsumer
dataConsumer: DataConsumer = DataConsumer@4616af54
```

9. For each line in the `BufferedReader`, pass it through the `accept` method of `DataConsumer`. This will cause all of the lines to be collected inside the `ListBuffer` of `dataConsumer`:

- Calling the lines method on the `reader` object returns all of the lines in the stream as an instance of `java.util.Stream`. The details of this can be found at `https://docs.oracle.com/javase/8/docs/api/java/util/stream/Stream.html`.
- The `forEach` construct of `java.util.Stream` provides a convenient way to apply the same operation to each element of the stream. In our case, each element is a line and we are collecting a `ListBuffer` object for each line:

```scala
scala> reader.lines().forEach(dataConsumer)
```

It is to be noted that the `forEach` construct used here is Java-specific. Scala has something similar, called `foreach`, which works quite differently from Java's `forEach` construct. In a similar situation, with Scala, `foreach`, we could have simply used the following:

```
reader.lines().foreach(i => buf += i) // buf is ListBuffer
```

Since we are using Java's `InputStream` family of classes here, we are forced to use the `forEach` construct of Java. Scala does not have equivalents of the Java Stream API.

10. Print out the first five elements of the `DataConsumer` class contained in `ListBuffer`:
    - The `DataConsumer` class's `buf` is a `ListBuffer` object. The `toList` method of `ListBuffer` returns a new `List` object consisting of all the elements in the buffer.
    - Take a convenient method of List to get the first specified number of elements as another List.
    - The `foreach` construct of List provides a mechanism to perform an action on each element. We are printing the element here:

```
scala> dataConsumer.buf.toList.take(5).foreach(println)
Zip Code,Total Population,Median Age,Total Males,Total
Females,Total Households,Average Household Size
91371,1,73.5,0,1,1,1
90001,57110,26.6,28468,28642,12971,4.4
90002,51223,25.5,24876,26347,11731,4.36
90003,66266,26.3,32631,33635,15642,4.22
```

In this example, the data producer has stored the data that can be extracted using an HTTP request. The data consumer has to explicitly issue the HTTP request in order to extract this data. This is a classic example of pull-based data extraction.

 Please note that the URL is using HTTPS, and this implies that data will be SSL/TLS encrypted in transit.

Using encryption for data in transit is very important to ensure that it does not get tampered with and accessed by someone, as in a man-in-the-middle security attack. On the same note, SFTP should always be preferred over FTP if a file transfer is to be used for exchanging data.

Push-oriented data delivery

Push-oriented data delivery reverses the data flow that was happening in a pull-oriented approach. In this case, the producer delivers that data to the consumer. Some of the most commonly used push mechanisms are as follows:

- **REST**: The consumer implements REST-based services that are invoked by the producer to publish new data, update existing data, or delete records.
- **Web services**: The consumer implements web services that are invoked by the producer in a fashion similar to the REST-based mechanism.
- **Pub-Sub**: The producer publishes the data to a generalized pub-sub framework, such as Kafka/message queue. The consumer subscribes to these messages and gets notified on arrival of new messages.
- **FTP/SFTP**: The FTP/SFTP mechanism is similar to a pull-oriented approach, however, the roles are reversed. In this case, the producer pushes the data to the consumer using the FTP/SFTP protocol. Many legacy systems used to deliver data using this mechanism.

This diagram illustrates how a **Data Producer** pushes data to **Data Consumer**:

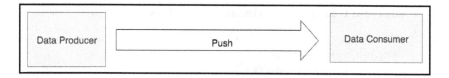

Let's look at a simple and concrete example of push-based data delivery. We will follow the REST API example provided by Play Framework (`https://github.com/playframework/play-scala-rest-api-example`). Play Framework is a powerful tool for building Scala and Java web applications. The documentation of this framework can be found at `https://www.playframework.com/`:

1. Clone the `play-scala-rest-api-example` repository (`https://github.com/playframework/play-scala-rest-api-example`) from GitHub. This needs to be cloned, because this project needs to be compiled. Run the `clone` command in a Terminal:

```
# clone from github
$ git clone
https://github.com/playframework/play-scala-rest-api-example
Cloning into 'play-scala-rest-api-example'...
...
Resolving deltas: 100% (385/385), done.
```

2. Change to the play example directory in the Terminal:

```
$ cd play-scala-rest-api-example
```

3. Run sbt by typing the following command in the Terminal:

```
$ sbt run
[info] Loading settings for project global-plugins from idea.sbt
...
...
--- (Running the application, auto-reloading is enabled) ---

[info] p.c.s.AkkaHttpServer - Listening for HTTP on
/0:0:0:0:0:0:0:0:9000

(Server started, use Enter to stop and go back to the console...)
```

HTTP REST services are now running at http://localhost:9000/. Let this server keep running in a Terminal while we explore further.

4. The REST end-point is http://localhost:9000/v1/posts. You can open this link in a web browser or run the following curl command to check the service:

```
$ curl curl http://localhost:9000/v1/posts
[
  {
    "id": "1",
    "link": "/v1/posts/1",
    "title": "title 1",
    "body": "blog post 1"
  },
  {
    "id": "2",
    "link": "/v1/posts/2",
    "title": "title 2",
    "body": "blog post 2"
  },
  {
    "id": "3",
    "link": "/v1/posts/3",
    "title": "title 3",
    "body": "blog post 3"
  },
  {
    "id": "4",
    "link": "/v1/posts/4",
    "title": "title 4",
    "body": "blog post 4"
```

```
    },
    {
      "id": "5",
      "link": "/v1/posts/5",
      "title": "title 5",
      "body": "blog post 5"
    }
```

The REST server created five sample blog post entries. Each blog post entry has the following attributes:

- id
- link
- title
- body

5. We can now POST some data to this REST server using the following curl command. We will now add the following new blog post entry:
   - id = 999
   - link = /v1/posts/999
   - title = mytitle
   - body = mybody

```
$ curl -d "title=mytitle&amp;amp;body=mybody" -X POST
http://localhost:9000/v1/posts
{"id":"999","link":"/v1/posts/999","title":"mytitle","body":"mybody"}
```

6. On the Terminal REST server, you should be seeing messages like these:

```
[trace] v.p.PostActionBuilder - invokeBlock:
 [trace] v.p.PostController - process:
     [trace] v.p.PostRepositoryImpl - create: data =
          PostData(999,mytitle,mybody)
```

In this example, we have pushed the data to the REST server using the curl command. The REST server, in this case, is the data consumer, and the curl command is the data producer.

One of the biggest benefits of push-oriented data delivery is that data is available to the consumer of the data in near real time. This is important, as there are several categories of data whose value rapidly diminishes with time and is generally of little value if not acted upon almost immediately.

Data staging

Once the data is extracted or delivered, it is generally staged into temporary storage for further processing. It is generally a good idea to keep data extraction/delivery storage separate from staging storage, although there are instances where this won't be necessary.

The staging area cleanly separates the following two aspects of the data ingestion process:

- Data that has been extracted or delivered
- New data that has to be processed

Once the data is staged completely, when we reach the further processing steps, such as cleaning, we do not have to be concerned about new data arriving. We can think of staged data as something that, once created, never changes and is immutable. This means that, to an already staged piece of data, no more data can be added, deleted, or modified. This is a very important data property that is necessary for reliable data processing in general. Any processor working on the staged data can safely assume that staged data will remain constant during the processing period.

Here, in this diagram, you can see the overall flow of data between the three key players:

- Data receiving area
- Data staging area
- Data processing

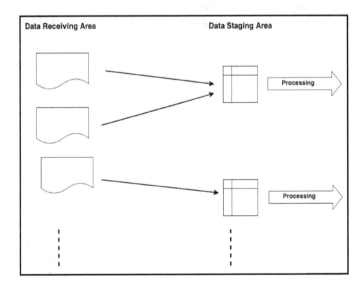

This is what is actually happening:

- Data is received or extracted into a **Data Receiving Area**. This could be as a result of pull-oriented data extraction or push-oriented data receipt.
- A set of data from the receiving area is moved into the **Data Staging Area**. This set of data in the staging area is immutable.
- Data processing takes place on the immutable dataset in the staging area. The data processing takes place only on the staged data.

Why is the staging important?

To answer this question, we have to understand a few realities on the ground. In an ideal world, the entire data processing works from end to end without any failures; however, in the real world, failures are quite common at different stages of processing. We need to account for these errors and develop a strategy for recovering from failures gracefully. In a large-scale data processing environment, we also want to recover from a safe and deterministic checkpoint so that the data processing that has already been done successfully can be reused, without repeating it. A staging area provides the unit of work that could be reliably used to achieve this goal.

We have three key participants here:

- **Staged-dataset-n**
- **Data processor**
- **Database**

Let's look at a simple example to illustrate this concept:

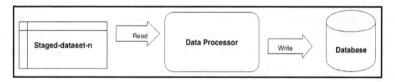

At a high level, the following is the flow:

- **Data Processor** reads the data from **Staged-dataset-n**
- **Data Processor** performs some transformations on the input data
- **Data Processor** writes the transformed data to the **Database**

So, the question is, how does staging help in this process? Let's assume that the data being written to the database is being tagged with the name **Staged-dataset-n**. Let's say that, in the middle of this data processing, the database server suddenly goes down. This will result in the data processor failing. When the database server comes back online, it has partial data from **Staged-dataset-n**. To recover from this failure, the following simple strategy could be followed:

1. Delete all of the data in the database associated with **Staged-dataset-n**
2. Rerun the **Data Processor** with **Staged-dataset-n**
3. Repeat this until **Staged-dataset-n** is successfully processed

This is a somewhat simple strategy to deal with unexpected failures during data processing; however, this illustrates a few important properties:

- This strategy is deterministic, because **Staged-dataset-n** is immutable. When the processing of this dataset fully completes, it will produce results consistent with the database.
- **Staged-dataset-n** is the unit of work, and the size of this unit of work can be controlled by the staging process. From the data processor's perspective, it has successfully processed this unit of work.

Cleaning and normalizing

Staged data is considered to be immutable. Immutability in this context implies that staged data, once created, never changes. Now, the data cleaning and normalizing process can start. This step could also involve determining the degree of errors in the data received. In particular, it is expected that big data will have a certain amount of errors.

Raw data coming from external sources comes in a variety of formats. These formats are generally designed for data delivery and are not suitable for use by systems consuming data. It is also very common for some of the information to be clubbed together as part of data delivery; however, the consumer of the data needs to have more fine-grained access to the information.

An example of this is the address part of the data. The data producer might provide a free-form address. The contained information, such as street name and city name, might also not be consistent across the dataset. On the other hand, it might be more desirable for the data consumer to have a high degree of confidence in the consistency of different components of the address, such as street name, city name, and ZIP code.

To perform data cleaning, it is often necessary to convert received data into an intermediate normal form first. The intermediate normal form is not necessarily the same as the final normal form, and is typically designed for efficient processing.

Let's consider a simple example of data related to persons in three different formats, namely XML, JSON, and CSV:

```
// XML
<person>
 <fname>Jon</fname>
 <lname>Doe</lname>
 <phone>123-456-7890</phone>
 <zip>12345</zip>
 <state>NY</state>
</person>

// JSON
{
 "person": {
 "fname": "Jon",
 "lname": "Doe",
 "phone": "123-456-7890",
 "zip": "12345",
 "state": "NY"
 }
}

// CSV
Jon,Doe,123-456-7890,12345,NY
```

Let's assume that there are several external data providers that provide this data; however, each provider chooses from one of the three formats. As a consumer of this data, you have to deal with all of these three formats depending upon the format of data provided by the producer. When data is being staged, it would make sense to create staged datasets that have one uniform data format.

The following could be the data processing flow:

- The staged dataset is one of three formats: XML, JSON, or CSV
- There is an intermediate step to determine the format of the staged dataset
- An appropriate data normalizer is selected based on the actual data format
- The staged dataset is normalized using the selected data normalizer
- The normalized data is processed by the data processor

Please note that, in this context, the normal form is still an intermediate form that must be able to capture all of the elements of incoming data, irrespective of the incoming data format (XML, JSON, or CSV).

The following diagram illustrates how all of this works together:

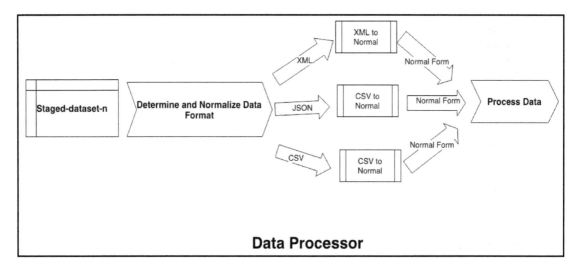

This is a very good example of a clear separation of concerns. The two concerns we are dealing with here are the following:

- Data coming in a variety of formats
- Processing of data

The processing of data does not depend upon the format of the data. It is dependent on the contents and associated meaning of that data. The normal form representation of data captures all of the elements of data without any loss of information. This provides greater flexibility when new data formats need to be supported. All that needs to be implemented are the following:

- A way to determine the new data format
- A converter that converts the new data format to normal form

The core processing remains unchanged because it is dependent upon the normal form only.

Let's explore this concept using a simple hands-on example using the **Scala Build Tool (SBT)**. Create a `build.sbt` class with the following info:

```
libraryDependencies ++= Seq(
// Scala XML library
  "org.scala-lang.modules" %% "scala-xml" % "1.1.0",
// Scala Lift JSON Library
  "org.json4s" %% "json4s-native" % "3.6.1"
  )
```

We are using two Scala libraries here:

- **Scala XML**: A powerful Scala library for processing the XML data format. The details of this project can found at `https://github.com/scala/scala-xml`.
- **Scala Lift JSON**: A powerful Scala library for processing the JSON data format. The details of this project can be found at `https://github.com/lift/framework/tree/master/core/json`.

We will explore how we can normalize XML, JSON, and CSV. For simplicity, we will be parsing CSV data using Scala's built-in `split` method of `String`:

1. Now, run SBT and start a Scala console. In the console explore, convert data in different formats to a normal form. We are using a Scala `case` class to represent a normal form.

2. Import `Elem` from Scala's `scala.xml` package:

   ```
   scala> import scala.xml.Elem
   import scala.xml.Elem
   ```

3. Import the following, and set `implicit`, which is needed for working with the JSON format:

   ```
   scala> import org.json4s._
   import org.json4s._

   scala> import org.json4s.native.JsonMethods._
   import org.json4s.native.JsonMethods._

   scala> implicit val formats = DefaultFormats
   formats: org.json4s.DefaultFormats.type =
                    org.json4s.DefaultFormats$@6f13a14e
   ```

4. Define a Scala `case` class called `Person` that will follow five attributes:
   - `fname`: First name
   - `lname`: Last name
   - `phone`: Phone number
   - `zip`: The ZIP code for where the person lives
   - `state`: The state where the person lives

```scala
scala> case class Person(fname: String, lname: String, phone:
                         String,
                         zip: String, state: String)
defined class Person
```

5. Create a JSON string that consists of the person's information, with all five attributes defined:

```scala
scala> // JSON

scala> val jsonStr = """
     | {
     | "fname": "Jon",
     | "lname": "Doe",
     | "phone": "123-456-7890",
     | "zip": "12345",
     | "state": "NY"
     | }
     | """
jsonStr: String =
"
{
    "fname": "Jon",
    "lname": "Doe",
    "phone": "123-456-7890",
    "zip": "12345",
    "state": "NY"
}
"
```

6. Parse a JSON string into a JSON object first:

```scala
scala> val json = parse(jsonStr)
json: org.json4s.JValue = JObject(List((fname,JString(Jon)),
      (lname,JString(Doe)), (phone,JString(123-456-7890)),
      (zip,JString(12345)), (state,JString(NY))))
```

7. Normalize the JSON object into Person:

```
scala> val normJson = json.extract[Person]
normJson: Person = Person(Jon,Doe,123-456-7890,12345,NY)
```

8. Create a Scala XML object:

```
scala> // XML

scala> val xml = <person>
     | <fname>Jon</fname>
     | <lname>Doe</lname>
     | <phone>123-456-7890</phone>
     | <zip>12345</zip>
     | <state>NY</state>
     | </person>
xml: scala.xml.Elem =
<person>
    <fname>Jon</fname>
    <lname>Doe</lname>
    <phone>123-456-7890</phone>
    <zip>12345</zip>
    <state>NY</state>
</person>
```

Please note that Scala automatically recognizes XML syntax and creates a `scala.xml.Elem` object.

9. Normalize the XML object into Person:

```
scala> val normXml = Person(xml \ "fname" text, xml \ "lname" text,
xml \ "phone" text, xml \ "zip" text, xml \ "state" text)

warning: there were 5 feature warnings; re-run with -feature for
details

normXml: Person = Person(Jon,Doe,123-456-7890,12345,NY)
```

Please note that Scala provides a convenient syntax to extract elements and attributes from a Scala XML `Elem` object. Here, https://github.com/scala/scala-xml/wiki/Getting-started is a good reference with which to become familiar with various features of Scala's XML capabilities.

10. Create a `CSV` object:

```
scala> // CSV (for simplicity, we use split method of String to
parse CSV)

scala> val csvStr = "Jon,Doe,123-456-7890,12345,NY"
csvStr: String = Jon,Doe,123-456-7890,12345,NY
```

11. Split the `CSV` object using the comma as a delimiter:

```
scala> val csvCols = csvStr.split(",")
csvCols: Array[String] = Array(Jon, Doe, 123-456-7890, 12345, NY)
```

12. Normalize `CSV` to `Person`:

```
scala> val normCsv = Person(csvCols(0), csvCols(1), csvCols(2),
                        csvCols(3), csvCols(4))
normCsv: Person = Person(Jon,Doe,123-456-7890,12345,NY)
```

13. Ensure that all three forms are the same:

```
scala> // Let us make sure that all three normal objects are same

scala> normXml == normJson
res0: Boolean = true

scala> normXml == normCsv
res1: Boolean = true

scala> normJson == normCsv
res2: Boolean = true
```

As can be seen, the normal form used previously is able to represent the data, arriving in different formats, into a single consistent structure. The remaining processing can rely on this form and can be completely agnostic toward what ever form the data came in.

In the preceding example, we assumed that the incoming data is perfectly clean; however, real-world data often requires some cleaning before it can be used. Let's assume that all of the data elements in `Person` cannot have leading or trailing spaces, and state must be capitalized. To perform this kind of data cleaning, we could do something like this:

1. Extend the functionality of the consumer case by adding a method called `cleanCopy`. This method returns a new `Person` instance with trimmed `fname`, `lname`, `phone`, `zip`, `state`, and `state` converted to upper case:

```
scala> case class Person(fname: String, lname: String, phone:
String, zip: String, state: String) {
```

```
    | def cleanCopy (): Person = {
    | this.copy(fname.trim, lname.trim, phone.trim, zip.trim,
              state.trim.toUpperCase)
    | }
    | }
```

```
defined class Person
```

2. Create a CSV record that is not cleaned:

```
scala> val uncleanCsvStr = " Jon , Doe , 123-456-7890 , 12345 , ny
"
    uncleanCsvStr: String = " Jon , Doe , 123-456-7890 , 12345 , ny
"
```

3. Split the CSV record with a comma as the delimiter:

```
scala> val uncleanCsvCols = uncleanCsvStr.split(",")
uncleanCsvCols: Array[String] = Array(" Jon ", " Doe ", " 123-456-
                          7890 ", " 12345 ", " ny ")
```

4. Create an unclean normalized `Person` object:

```
scala> val uncleanNormCsv = Person(uncleanCsvCols(0),
uncleanCsvCols(1), uncleanCsvCols(2), uncleanCsvCols(3),
uncleanCsvCols(4))
uncleanNormCsv: Person = Person( Jon , Doe , 123-456-7890 , 12345 ,
ny )
```

5. Create a clean, normalized `Person` object from an unclean one:

```
scala> val cleanNormCsv = uncleanNormCsv.cleanCopy
cleanNormCsv: Person = Person(Jon,Doe,123-456-7890,12345,NY)
```

Let's revisit the `Person` case class and the `cleanCopy` extension that was made earlier. Inside our `Person` case class, we defined a new method that returns a cleaned copy of the `Person` object. Scala case classes provide a method called `copy`, which can be used to create a copy of the same object, but optionally modifies a subset of attributes as needed. In our case, we have modified all of the attributes of the original. We could have constructed a new `Person` object instead of using `copy`:

```
Person(fname.trim, lname.trim, phone.trim, zip.trim,
       state.trim.toUpperCase)
```

The following `Person` case class implements the same functionality as the earlier one:

```
case class Person(fname: String, lname: String, phone: String, zip: String,
state: String) {
  def cleanCopy(): Person = {
    Person(fname.trim, lname.trim, phone.trim, zip.trim,
state.trim.toUpperCase)
  }
}
```

The preceding example is quite simple, but illustrates that simple data cleaning can be performed with ease using built-in Scala features, and Scala case classes really come in handy when working with normal forms of data.

Enriching

Data enrichment is the act of adding more information to raw data. Some examples of the enrichments are as follows:

- Adding missing values
- Adding lookup values
- Joining with other datasets
- Filtering
- Aggregating

Continuing on the same example of using `Person`, let's say that the `state` element is optional. Given the `zip` information, we should be able to derive the value of `state`. In this specific case, we are performing the following two enrichments:

- Looking up or deriving the value of `state` based on the `zip` value
- Adding missing `state` value

Let's define a simple Scala function to map a US zip code to `state`. Please refer to Wikipedia (`https://en.wikipedia.org/wiki/List_of_ZIP_code_prefixes`) for more info on US ZIP codes:

1. Define a Scala case class called `Person` with a method called `cleanCopy` to provide a clean copy of the object:

```
scala> case class Person(fname: String, lname: String, phone:
String, zip: String, state: String) {
    | def cleanCopy(): Person = {
```

```
      | this.copy(fname.trim, lname.trim, phone.trim, zip.trim,
state.trim.toUpperCase)
      | }
      | }
```

```
defined class Person
```

2. Define a function called `getState` that finds a US state when given a US ZIP code:

  - **Input:** `zip`
  - **Output:** `state`

Please note that this only a partial implementation of this function:

```
scala> def getState(zip: String): String = { // Partial implementation for
simplicity
 | val zipPerfix = zip.substring(0, 3)
 | zipPerfix match {
 | case "006"|"007"|"009" => "PR" // PR is Puerto Rico
 | case "008" => "VI" // VI is Virgin Islands
 | case n if (n.toInt >= 10 &amp;amp;&amp;amp; n.toInt <= 27) => "MA" //
010 to 027 is MA
 | case "028" | "029" => "RI"
     | case n if (n.toInt >= 100 &amp;amp;&amp;amp; n.toInt <= 149) => "NY"
// 010 to 027 is MA
     | case _ => "N/A"
     | }
     | }
getState: (zip: String)String
```

3. Define a function called `populateStateIfNecessary` that populates the `state` attribute of `Person`, if it is not already populated. It makes use of `getState` function to find state based on the ZIP code:

  - **Input:** `Person`
  - **Output:** `Person`

```
scala> def populateStateIfNecessary(p: Person): Person = {
    | if (p.state == null || p.state.isEmpty)
    | p.copy(state=getState(p.zip))
    | else
    | p
    | }
populateStateIfNecessary: (p: Person)Person
```

Please note that preceding function returns the same `Person` object of the `state` attribute if already populated. If it is not populated, it creates a copy of the `Person` object, sets the `state` attribute, of copy to the `state` derived using the ZIP code, and then returns this copy of the `Person` object.

 Here is an important tip: creating unnecessary objects should be avoided as far as possible. This is because these objects occupy memory space in JVM's heap space. Another side effect of this could be that there are frequent Java **Garbage Collection** (**GC**) activities happening in JVM, which lead to degradation in the overall performance of the application.

4. Create an unclean CSV record with the missing `state` information:

```scala
scala> val uncleanCsvStr = " Jon , Doe , 123-456-7890 , 12345 , "
// missing state
uncleanCsvStr: String = " Jon , Doe , 123-456-7890 , 12345 , "
```

5. Split the unclean CSV record by using the comma as the delimiter:

```scala
scala> val uncleanCsvCols = uncleanCsvStr.split(",")
uncleanCsvCols: Array[String] = Array(" Jon ", " Doe ", "
123-456-7890 ", " 12345 ", " ")
```

6. Create an unclean normal object from the unclean CSV record:

```scala
scala> val uncleanNormCsv = Person(uncleanCsvCols(0),
uncleanCsvCols(1), uncleanCsvCols(2), uncleanCsvCols(3),
uncleanCsvCols(4))
uncleanNormCsv: Person = Person( Jon , Doe , 123-456-7890 , 12345 ,
)
```

7. Clean the unclean normal record:

```scala
scala> val cleanNormCsv = uncleanNormCsv.cleanCopy
cleanNormCsv: Person = Person(Jon,Doe,123-456-7890,12345,)
```

8. Enrich the normal record:

```scala
scala> val enriched = populateStateIfNecessary(cleanNormCsv)
enriched: Person = Person(Jon,Doe,123-456-7890,12345,NY)
```

The preceding example illustrates how to enrich data. We can also think of the aforementioned example, similar to value lookup, which can be implemented as a dataset join operation with a lookup dataset. In this case, the lookup data has a mapping from ZIP code to state. Every single ZIP value needs to be mapped to the state for this to work.

There are times when incoming data has more data than the data consumer is interested in.

We can filter out the unnecessary data and only keep that data we are interested in:

1. Create input data that has a mix of useful, as well as some unnecessary, records:

```scala
scala> val originalPersons = List(
     | Person("Jon","Doe","123-456-7890","12345","NY"),
     | Person("James","Smith","555-456-7890","00600","PR"),
     | Person("Don","Duck","777-456-7890","00800","VI"),
     | Person("Doug","Miller","444-456-7890","02800","RI"),
     | Person("Van","Peter","333-456-7890","02700","MA")
     | )
originalPersons: List[Person] =
List(Person(Jon,Doe,123-456-7890,12345,NY),
Person(James,Smith,555-456-7890,00600,PR),
Person(Don,Duck,777-456-7890,00800,VI),
Person(Doug,Miller,444-456-7890,02800,RI),
Person(Van,Peter,333-456-7890,02700,MA))
```

2. Define the exclusion states:

```scala
scala> val exclusionStates = Set("PR", "VI") // we want to exclude
these states
exclusionStates: scala.collection.immutable.Set[String] = Set(PR,
VI)
```

3. Filter out the records belonging to exclusion states:

```scala
scala> val filteredPersons = originalPersons.filterNot(p =>
exclusionStates.contains(p.state))
filteredPersons: List[Person] =
List(Person(Jon,Doe,123-456-7890,12345,NY),
Person(Doug,Miller,444-456-7890,02800,RI),
Person(Van,Peter,333-456-7890,02700,MA))
```

In the preceding example, we wanted to exclude data from certain states, since that is not relevant to our data analysis. The technique demonstrated appears fairly simple because of powerful constructs provided by the Scala programming language. The `filterNot` method of the Scala collection API removes any element from the collection that satisfies the condition being tested.

Alternatively, we could have used the `filter` API, which is inverse of the `filterNot` API. Let's see this in action:

1. Create a list of mixed records:

```scala
scala> val originalPersons = List(
     | Person("Jon","Doe","123-456-7890","12345","NY"),
     | Person("James","Smith","555-456-7890","00600","PR"),
     | Person("Don","Duck","777-456-7890","00800","VI"),
     | Person("Doug","Miller","444-456-7890","02800","RI"),
     | Person("Van","Peter","333-456-7890","02700","MA")
     | )
originalPersons: List[Person] =
List(Person(Jon,Doe,123-456-7890,12345,NY),
Person(James,Smith,555-456-7890,00600,PR),
Person(Don,Duck,777-456-7890,00800,VI),
Person(Doug,Miller,444-456-7890,02800,RI),
Person(Van,Peter,333-456-7890,02700,MA))
```

2. Define exclusion states:

```scala
scala> val exclusionStates = Set("PR", "VI") // we want to exclude
these states
exclusionStates: scala.collection.immutable.Set[String] = Set(PR,
VI)
```

3. Use the `filterNot` API first to remove unwanted records:

```scala
scala> val filteredPersons1 = originalPersons.filterNot(p =>
exclusionStates.contains(p.state))
filteredPersons1: List[Person] =
List(Person(Jon,Doe,123-456-7890,12345,NY),
Person(Doug,Miller,444-456-7890,02800,RI),
Person(Van,Peter,333-456-7890,02700,MA))
```

4. Use the `filter` API next to remove unwanted records:

```scala
scala> val filteredPersons2 = originalPersons.filter(p =>
!exclusionStates.contains(p.state))
filteredPersons2: List[Person] =
List(Person(Jon,Doe,123-456-7890,12345,NY),
Person(Doug,Miller,444-456-7890,02800,RI),
Person(Van,Peter,333-456-7890,02700,MA))
```

5. Compare the two results:

```scala
scala> filteredPersons1 == filteredPersons2
res2: Boolean = true
```

You can see that we can produce identical outcomes using the `filter` and `filterNot` APIs by just inverting the test conditions.

 Please note that, in the real-world applications, you would notice that the `filter` API is used significantly more than the `filterNot`. It is generally a matter of personal preference at times; however, code readability should be the primary criteria in making an appropriate choice between the two.

Organizing and storing

Once all of the processing related to data is completed, it is organized and stored in such a way that consumers of data can begin using the data. This act is equivalent to publishing data and making it available for usage.

Data usage generally drives how data is organized and stored. Time series data is generally organized by time dimensions. For example, we could organize data coming from **Internet of Things (IoT)** devices in the following hierarchy:

- Year
- Month
- Day
- Hour
- Minute

In this context, year, month, day, hour, and minute are the functions of the event occurrence time. For example, if the occurrence time is 2018-10-31 15:30:45 UTC, then it will have the following values:

- Year = 2018
- Month = 10
- Day = 31
- Hour = 15
- Minute = 30

Another way to organize the data could be by using the device type, or even a mix of both time dimensions and device type. In the context of big data, it is often the case that multiple storage strategies are devised to store multiple copies of the same data, but different storage structures are used to support use cases where a single structure is unable to support all of the use cases efficiently and effectively.

A significant number of factors need to be considered when it comes to data organization and storage. The data model plays the most central role when the end user's perspective is taken into consideration. It also needs to be considered how the security and retention policies around this data will be handled. All of these factors contribute to determining the appropriate strategy for managing this data. We essentially have three distinct forces at work:

- The raw data
- The processed and organized data
- The data users

The following is a diagram that illustrates the interaction between these players at a high level:

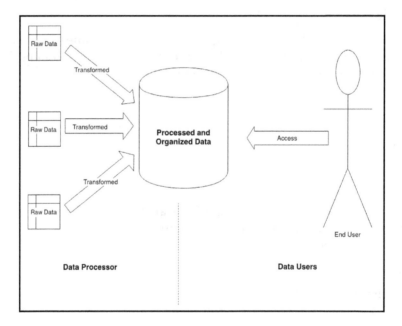

The preceding diagram illustrates the data's journey from raw to organized and its usage interactions by end users of the data. This highlights the fact that processed and organized data should be able to answer end users' questions effectively and efficiently. This should be the primary focus for determination of storage and organizational structure of the processed data.

It is not necessary that the end user is a human being. The data user could even be another computer process. A significant effort is generally devoted to coming up with an appropriate data model to support a variety of end users. In the big data world, it is also quite common to have multiple data models for the same dataset to support fairly diverse use cases.

Summary

In this chapter, we looked at some of the key tasks associated with aspects with data ingestion. In particular, we looked at data staging and dealing with various data formats. We also got an understanding of data cleaning and enrichment. We also looked at how to organize and store data so that it can be used for data analysis. The end user's or consumer's perspective is very important when it comes to defining an appropriate data model, and all of the important use cases must be taken into consideration. At times, there is a need to create multiple data models to support use cases that have completely different needs.

In the next chapter, we will look at exploring and visualizing data.

4
Data Exploration and Visualization

Data exploration is about trying to gain an understanding of patterns and relationships hidden inside the data. Data visualization helps tremendously in this process. In fact, visual methods are frequently used to explain and communicate these patterns and relationships to an interested audience. It needs to be noted that data exploratory analysis and data explanatory analysis are two different things. Data explanatory analysis can only start after data exploratory analysis is completed. Our focus here is primarily data exploratory analysis and we want to discover and learn about the structure of data. Visual tools play a more dominant role in explanatory data analysis; however, these also play an equally important role during data exploration.

The following are the topics that we will be covering in this chapter:

- Sampling data
- Performing ad hoc analysis
- Finding a relationship between data elements
- Visualizing data

Sampling data

To explore large datasets, it is generally useful to work with a smaller sample of data first. For example, from a dataset consisting of 100 million records, we could take a sample of 1,000 records and start exploring some important properties of this data. Exploring the entire dataset would be ideal; however, the time required to do so would increase manifold.

Selecting the sample

For working with samples, it is important that sample selection is done carefully and biases are not introduced unnecessarily. Randomness plays a very important role in this.

Let's look at how we can make use of the Scala collection API to select sample data from a dataset:

1. Create a list of 1000 numbers using Scala's Range API. We generate a sequence of 1,000 number from 0 to 1000 (1,000 is excluded) first and turn it into a Scala List:

```
scala> val data = Range(0, 1000).toList
data: List[Int] = List(0, 1, 2, 3, 4, 5, 6, 7, 8, 9, 10, 11, 12,
13, 14, 15, 16, 17, 18, 19, 20, 21, 22, 23, 24, 25, 26, 27, 28, 29,
30, 31, 32, 33, 34, 35, 36, 37, 38, 39, 40, 41, 42, 43, 44, 45, 46,
47, 48, 49, 50, 51, 52, 53, 54, 55, 56, 57, 58, 59, 60, 61, 62, 63,
64, 65, 66, 67, 68, 69, 70, 71, 72, 73, 74, 75, 76, 77, 78, 79, 80,
81, 82, 83, 84, 85, 86, 87, 88, 89, 90, 91, 92, 93, 94, 95, 96, 97,
98, 99, 100, 101, 102, 103, 104, 105, 106, 107, 108, 109, 110, 111,
112, 113, 114, 115, 116, 117, 118, 119, 120, 121, 122, 123, 124,
125, 126, 127, 128, 129, 130, 131, 132, 133, 134, 135, 136, 137,
138, 139, 140, 141, 142, 143, 144, 145, 146, 147, 148, 149, 150,
151, 152, 153, 154, 155, 156, 157, 158, 159, 160, 161, 162, 163,
164, 165, 166, 167, 168, 169, 170, 171, 172, 173, 174, 175, 176,...
```

2. Use the Scala list's take method, select the first three elements of the aforementioned-generated List. This will provide another List with three elements:

```
scala> val first3 = data.take(3)
first3: List[Int] = List(0, 1, 2)
```

We generated a list of 1,000 integers from 0 to 999 and selected the first three integers from this. The previous steps would always produce the same result each time.

This implies that if the dataset remains constant, then the selected values would always be the same and our sample has implicit bias. Let's see how we can select random values:

1. Import the Random class from Scala's util package:

```
scala> import scala.util.Random
import scala.util.Random
```

2. Perform a random shuffle operation on the previously generated data, using Scala's random utility class's `shuffle` method. This produces another list with the same content as the original one; however, the position of the numbers in the new list is randomized and different from the original list:

```scala
scala> val randomizedData = Random.shuffle(data)
randomizedData: List[Int] = List(725, 225, 231, 280, 518, 818, 395,
519, 13, 648, 292, 826, 520, 885, 114, 403, 277, 218, 707, 864,
798, 575, 942, 685, 627, 95, 512, 753, 763, 923, 209, 633, 631,
743, 327, 0, 946, 147, 838, 78, 777, 473, 521, 501, 86, 590, 748,
956, 105, 963, 483, 334, 109, 5, 285, 910, 791, 102, 398, 240, 447,
493, 351, 297, 399, 365, 466, 612, 298, 529, 762, 680, 975, 253,
535, 902, 373, 36, 356, 596, 679, 717, 976, 543, 180, 894, 500,
624, 405, 754, 881, 916, 213, 768, 305, 740, 263, 422, 771, 623,
121, 989, 486, 574, 196, 987, 968, 73, 943, 662, 393, 438, 834,
714, 746, 364, 260, 139, 906, 944, 793, 485, 647, 66, 418, 909,
787, 377, 94, 25, 84, 888, 657, 23, 776, 402, 649, 472, 915, 496,
140, 155, 772, 319, 752, 964, 354, 11, 431, 413, 982, 621, 835,
468, 785, 463, ...
```

3. Select the first three elements from the randomized list, using the `take` method:

```scala
scala> val random3 = randomizedData.take(3)
random3: List[Int] = List(725, 225, 231)
```

4. Repeat step 2 to produce another randomized list. This randomized list is expected to be quite different compared to the original list and previously generated list:

```scala
scala> val randomizedDataNext = scala.util.Random.shuffle(data)
randomizedDataNext: List[Int] = List(955, 128, 857, 129, 901, 265,
535, 879, 998, 373, 601, 816, 297, 648, 624, 27, 119, 195, 868,
357, 859, 986, 569, 660, 167, 885, 416, 199, 848, 406, 751, 593,
156, 673, 333, 403, 628, 122, 775, 390, 926, 360, 513, 953, 820,
947, 867, 295, 113, 639, 897, 856, 717, 426, 865, 988, 407, 814,
110, 762, 852, 842, 940, 102, 61, 298, 815, 197, 233, 515, 318,
401, 180, 781, 262, 157, 492, 376, 747, 688, 186, 824, 961, 659,
269, 618, 819, 623, 866, 46, 557, 511, 176, 840, 800, 679, 481,
704, 551, 66, 54, 977, 732, 700, 813, 264, 625, 171, 347, 990, 290,
43, 742, 418, 836, 92, 979, 938, 369, 111, 779, 3, 613, 117, 379,
8, 764, 356, 573, 921, 893, 822, 351, 279, 164, 507, 930, 514, 805,
245, 714, 121, 694, 223, 652, 526, 755, 692, 260, 476, 105, 404,
289, 869, 5...
```

5. Select the first three elements from the new randomized `List` using the `take` method:

```
scala> val random3Next = randomizedDataNext.take(3)
random3Next: List[Int] = List(955, 128, 857)
```

In this code example, we are to able to get three random values by using the `scala.util.Random.shuffle` function. Although the preceding example illustrated the data randomization technique, it is not very efficient in terms of performance and it won't scale as the datasets get larger and larger. It does, however, illustrate a simple way to get random samples using Scala's built-in APIs. We will look at how to efficiently get random samples from large datasets in subsequent chapters.

The details of Scala's random utility can be found at `https://www.scala-lang.org/api/2.12.8/scala/util/Random$.html`. Some of the commonly used APIs are shown in the following table:

Method name	Purpose
nextInt	Provides the next pseudo-random uniformly distributed integer
nextLong	Provides the next pseudo-random uniformly distributed long
nextDouble	Provides the next pseudo-random uniformly distributed double between 0.0 and 2.0
nextBytes	Fills the user-supplied array of bytes with the next set of pseudo-random uniformly distributed bytes
nextGaussion	Provides the next pseudo-random Gaussian distributed double value with a mean of 0.0 and a standard deviation of 1.0
nextPrintableChar	Provides the next pseudo-random uniformly distributed printable character from the ASCII character set
shuffle	Requires a collection as input and provides a new collection whose element positions are randomized

Selecting samples using Saddle

Let's look at a similar exercise using the Scala Saddle library. We will be using the CSV data from `https://data.lacity.org/api/views/nxs9-385f/rows.csv?accessType=DOWNLOAD`.

This dataset was introduced in earlier chapters. Let's follow these steps to use Saddle:

1. First, we need to define our `build.sbt`, as follows, to include the Saddle library dependencies. Remember to save `build.sbt` as a file in your current directory:

```
scalaVersion := "2.11.12"

libraryDependencies ++= Seq(
  "org.scala-saddle" %% "saddle-core" % "1.3.4" // Saddle Dataframe
like Library
)
```

2. Start SBT in your Terminal from the same directory where `build.sbt` is located and start a Scala console:

```
$ sbt
```

3. Import `BufferedReader` and `InputStreamReader` from the `java.io` package:

```
scala> import java.io.{BufferedReader, InputStreamReader}
import java.io.{BufferedReader, InputStreamReader}
```

4. Import the `saddle` package:

```
scala> import org.saddle.io._
import org.saddle.io._
```

5. Define a Scala class called `SaddleCsvSource` that takes a URL string as an input argument to the constructor and extends `CsvSource`. The constructor establishes a connection to the provided URL and creates a `BufferedReader` object that can be used to read data from the URL, line by line:

```
scala> class SaddleCsvSource(url: String) extends CsvSource {
     | val reader = new BufferedReader(new InputStreamReader(new
java.net.URL(url).openStream()))
     | override def readLine: String = {
     | reader.readLine()
     | }
     | }
defined class SaddleCsvSource
```

We have overridden the `readLine` method of the parent `CsvSource` class. The overridden method reads a line of data from the URL. This method automatically gets repeated when the parse method is invoked on `CsvSource`.

6. Create a new instance of `SaddleCsvSource` by supplying `https://data.lacity.org/api/views/nxs9-385f/rows.csv?accessType=DOWNLOAD` as the URL. This is our source data that is in CSV format and we want to parse this data using Saddle's CSV parser:

```scala
scala> val file = new
SaddleCsvSource("https://data.lacity.org/api/views/nxs9-385f/rows.c
sv?accessType=DOWNLOAD")
file: SaddleCsvSource = SaddleCsvSource@3f0055eb
```

7. Parse the aforementioned object using the `CsvParser` instance parse, API. This provides Saddle's `Frame` object that is used for further exploration:

```scala
scala> val frameOrig = CsvParser.parse(file)
frameOrig: org.saddle.Frame[Int,Int,String] =
[320 x 7]
              0 1 2 3 4 5 6
      -------- ---------------- ---------- ----------- -----------
-- ---------------- ---------------------
  0 -> Zip Code Total Population Median Age Total Males Total
Females Total Households Average Household Size
  1 -> 91371 1 73.5 0 1 1 1
  2 -> 90001 57110 26.6 28468 28642 12971 4.4
  3 -> 90002 51223 25.5 24876 26347 11731 4.36
  4 -> 90003 66266 26.3 32631 33635 ...
scala>
```

8. Get the header:

```scala
scala> val head = frameOrig.rowSlice(0,1).rowAt(0)
head: org.saddle.Series[Int,String] =
[7 x 1]
0 -> Zip Code
1 -> Total Population
2 -> Median Age
3 -> Total Males
4 -> Total Females
5 -> Total Households
6 -> Average Household Size
scala>
```

9. Remove the header row and attach the header back as column names:

```
scala> val frame = frameOrig.rowSlice(1,
frameOrig.numRows).mapColIndex(i => head.at(i).get)
frame: org.saddle.Frame[Int,String,String] =
[319 x 7]
      Zip Code Total Population Median Age Total Males Total
Females Total Households Average Household Size
      -------- ---------------- ---------- ----------- -----------
-- ---------------- ----------------------
   1 -> 91371 1 73.5 0 1 1 1
   2 -> 90001 57110 26.6 28468 28642 12971 4.4
   3 -> 90002 51223 25.5 24876 26347 11731 4.36
   4 -> 90003 66266 26.3 32631 33635 15642 4.22
   5 -> 90004 62180 34.8 31302 30878 2...
```

10. Get the first three records from Saddle's `Frame`:

```
scala> frame.head(3)
res1: org.saddle.Frame[Int,String,String] =
[3 x 7]
      Zip Code Total Population Median Age Total Males Total Females
Total Households Average Household Size
      -------- ---------------- ---------- ----------- -------------
---------------- ----------------------
1 -> 91371 1 73.5 0 1 1 1
2 -> 90001 57110 26.6 28468 28642 12971 4.4
3 -> 90002 51223 25.5 24876 26347 11731 4.36
scala>
```

11. Get a random sample of 2% of the dataset by using Frame's `rfilter` API. Note the usage of Scala's random utility's `nextDouble` method. This method provides a uniformly distributed pseudo-random double between 0.0 and 1.0. This implies that roughly only 2% of the time the following condition will hold true if called repeatedly:

- `scala.utilRandom() < 0.02`

The `rfilter`, when combined with this mechanism, provides us with roughly 2% of the sample data:

```
scala> val sample = frame.rfilter(_ => scala.util.Random.nextDouble() <
0.02)
sample: org.saddle.Frame[Int,String,String] =
[6 x 7]
      Zip Code Total Population Median Age Total Males Total Females Total
Households Average Household Size
-------- ---------------- ----------- ----------- ------------- --------
```

```
  7 -> 90006 59185 32.4 30254 28931 18617 3.13
 80 -> 90241 42399 33.9 20466 21933 13617 3.09
126 -> 90606 32396 33.5 15936 16460 8633 3.72
156 -> 90802 39347 34.7 20387 18960 19853 1.93
259 -> 91722 34409 34 16859 17550 10...
```

We are now able to get a sample of data conveniently using the APIs provided by the Saddle library.

Please note the abstraction being used in this context is a rame that consists of rows and columns. It is to be noted that every run of a sample would produce a different result. For example:

```
scala> val sample = frame.rfilter(_ => scala.util.Random.nextDouble() <
0.02)
sample: org.saddle.Frame[Int,String,String] =
[10 x 7]
      Zip Code Total Population Median Age Total Males Total Females Total
Households Average Household Size
      -------- ----------------- ---------- ----------- ------------- -----
----------- ----------------------
 17 -> 90017 23768 29.4 12818 10950 9338 2.53
 18 -> 90018 49310 33.2 23770 25540 15493 3.12
 54 -> 90062 32821 31.8 15720 17101 9155 3.55
 90 -> 90262 69745 27.8 33919 35826 14669 4.57
101 -> 90290 6368 45 3180 3188 2612 2.44
131 -> 90638 49012 37.9 23520 25492 14821 3.11
188 -> 91108 13361 45.4 6410 6951 4415 3.01
273 -> 91755 27496 43.4 13271 14225 8760 3.12
281 -> 91773 33119 42.5 15737 17382 11941 2.73
311 -> 93543 13033 32.9 6695 6338 3560 3.66
```

Although the samples are small, these, however, provide an important insight into some of the properties of the data, such as typical median age and average household size. We performed the following activities in our analysis:

1. We started with a data resource on the internet, located at https://data. lacity.org/api/views/nxs9-385f/rows.csv?accessType=DOWNLOAD. This resource is in CSV format.
2. Using a combination of Java's and Saddle's APIs, we were able to read this dataset.

3. Saddle's API allowed us to parse the CSV data and convert this into a structured format of Saddle's frame.

4. Saddle's frame allowed us to see the source data in a tabular form, consisting of rows and columns.

5. We conveniently got a sample, but a randomized set of rows from the frame, by using the `rfilter` API and combining it with Scala's random utility's `nextDouble` API.

Performing ad hoc analysis

We can use ad hoc analysis to learn about important properties of the data. Some of the issues that can be easily solved with the data are:

- Statistical properties, such as mean, median, the range for numerical data
- Distinct values for numerical as well as non-numerical data
- The frequency of data occurrence

We can ask these questions on a sample of data or an entire dataset. With a distributed framework, such as Spark, it is quite easy and convenient to get answers to these questions. In fact, many of these frameworks have a simple API to support this. Ad hoc analysis can also be performed on the very raw data itself. In this case, some of the data transformations are applied as part of the process. The main purpose of the ad hoc analysis is to gain a quick understanding of some of the properties of the data.

We will use Spark to perform some hands-on ad hoc analysis. Let's create an IntelliJ project with the following Spark dependencies added to `build.sbt`:

```
scalaVersion := "2.11.12"

libraryDependencies ++= Seq(
  "org.apache.spark" %% "spark-sql" % "2.4.0", // Spark Core Library
  "org.scalatest" %% "scalatest" % "3.0.5" % "test" // Scala test library
)
```

This is what it will look like:

Now, create a simple main Scala object to explore the same dataset that we explored in the previous section:

1. Import `SparkSession` from Spark's `sql` package. This is needed for setting up a Spark Session programmatically:

   ```
   import org.apache.spark.sql.SparkSession
   ```

2. Import `SparkFiles` from Spark's `spark` package. This is needed for reading data from the internet resource located at `https://data.lacity.org/api/views/nxs9-385f/rows.csv`:

   ```
   import org.apache.spark.SparkFiles
   ```

3. Define a Scala object called `SparkExample`, which will act an entry point into this program:

   ```
   object SparkExample {
   ...
   }
   ```

4. With the `SparkExample` object, define a method called `getSparkSession`. This creates a new Spark session running in local mode and uses the builder pattern to create or get a Spark session. The advantage of using the builder pattern is that the session being created can be customized to serve very specialized needs:

```
def getSparkSession(): SparkSession = {
    val spark =
SparkSession.builder().master("local").getOrCreate()
    spark.sparkContext.setLogLevel("ERROR")
    spark
}
```

5. Finally, define the main method that facilitates the program to be executed with input arguments supplied as parameters. There are multiple actions taking place inside this method:

   1. The Spark session is established by calling the `getSparkSession` method of the object.

   2. The internet resource located at `https://data.lacity.org/api/views/nxs9-385f/rows.csv` is being added to the `SparkContext` object of `spark`, the Spark session.

   3. A new Spark DataFrame is being created by fetching the internet resource and treating it as CSV data with a header row. Also, the schema of the target DataFrame is determined by inferring the contents of CSV.

   4. We print the schema of the DataFrame created.

   5. We show a few rows from the DataFrame.

   6. Stop the Spark Session on completion of the program.

```
def main(args: Array[String]): Unit = {
    val spark = getSparkSession()
spark.sparkContext.addFile("https://data.lacity.org/api/views/n
xs9-385f/rows.csv")
    val df = spark.read.option("header",
true).option("inferSchema",
true).csv(SparkFiles.get("rows.csv"))
    df.printSchema()
    df.show()
    spark.stop()
}
```

Let's put all of this together as a single program. The complete code is outlined as follows:

```
import org.apache.spark.sql.SparkSession
import org.apache.spark.SparkFiles

object SparkExample {
  def getSparkSession(): SparkSession = {
    val spark = SparkSession.builder().master("local").getOrCreate()
    spark.sparkContext.setLogLevel("ERROR")
    spark
  }
  def main(args: Array[String]): Unit = {
    val spark = getSparkSession()
spark.sparkContext.addFile("https://data.lacity.org/api/views/nxs9-385f/row
s.csv")
    val df = spark.read.option("header", true).option("inferSchema",
true).csv(SparkFiles.get("rows.csv"))
    df.printSchema()
    df.show()
    spark.stop()
  }
}
```

When you run the preceding example in IntelliJ, it produces quite a bit of log information. It should output some logs like these:

- Pay attention to the schema Information. Note that the data types of the columns are inferred from the source data. Without the schema inference, the types of all the columns would have been string type:

```
root
 |-- Zip Code: integer (nullable = true)
 |-- Total Population: integer (nullable = true)
 |-- Median Age: double (nullable = true)
 |-- Total Males: integer (nullable = true)
 |-- Total Females: integer (nullable = true)
 |-- Total Households: integer (nullable = true)
 |-- Average Household Size: double (nullable = true)
```

- The output from the show method could look something like this. By default, the show method of the DataFrame display is 20 rows:

```
+--------+----------------+----------+-----------+------------+---
-
|Zip Code|Total Population|Median Age|Total Males|Total
Females|Total Households|Average Household Size|
+--------+----------------+----------+-----------+------------+---
```

```
-
| 91371|      1| 73.5|     0|    1|     1| 1.0|
| 90001|  57110| 26.6| 28468| 28642| 12971| 4.4|
| 90002|  51223| 25.5| 24876| 26347| 11731| 4.36|
| 90003|  66266| 26.3| 32631| 33635| 15642| 4.22|
| 90004|  62180| 34.8| 31302| 30878| 22547| 2.73|
| 90005|  37681| 33.9| 19299| 18382| 15044| 2.5|
| 90006|  59185| 32.4| 30254| 28931| 18617| 3.13|
| 90007|  40920| 24.0| 20915| 20005| 11944| 3.0|
| 90008|  32327| 39.7| 14477| 17850| 13841| 2.33|
| 90010|   3800| 37.8|  1874|  1926|  2014| 1.87|
| 90011| 103892| 26.2| 52794| 51098| 22168| 4.67|
| 90012|  31103| 36.3| 19493| 11610| 10327| 2.12|
| 90013|  11772| 44.6|  7629|  4143|  6416| 1.26|
| 90014|   7005| 44.8|  4471|  2534|  4109| 1.34|
| 90015|  18986| 31.3|  9833|  9153|  7420| 2.45|
| 90016|  47596| 33.9| 22778| 24818| 16145| 2.93|
| 90017|  23768| 29.4| 12818| 10950|  9338| 2.53|
| 90018|  49310| 33.2| 23770| 25540| 15493| 3.12|
| 90019|  64458| 35.8| 31442| 33016| 23344| 2.7|
| 90020|  38967| 34.6| 19381| 19586| 16514| 2.35|
+--------+---------------+----------+-----------+-------------+---
-

only showing top 20 rows
```

Hence, we were able to do a quick ad hoc analysis on this information to understand some of the properties of the data and examine a few sample records. The only assumption we made about the data is that its format is CSV and the first record is a header record. Using Spark, we are also able to infer the schema of the underlying data with appropriate data types.

Spark has a comprehensive API for ad hoc analysis. For example, to get random samples, we could do the following to get a 5% sample of rows from the entire DataFrame:

```
df.sample(0.05).show() // 5% random sample

+--------+---------------+----------+-----------+-------------+-------
|Zip Code|Total Population|Median Age|Total Males|Total Females|Total
Households|Average Household Size|
+--------+---------------+----------+-----------+-------------+-------
| 90002|  51223| 25.5| 24876| 26347| 11731| 4.36|
| 90039|  28514| 38.8| 14383| 14131| 11436| 2.47|
| 90073|    539| 56.9|   506|    33|     4| 1.25|
| 90201| 101279| 27.8| 50658| 50621| 24104| 4.16|
| 90263|   1612| 19.7|   665|   947|     0| 0.0|
| 90703|  49399| 43.9| 23785| 25614| 15604| 3.16|
| 90822|    117| 63.9|   109|     8|     2| 4.5|
```

```
|  91042|  27585|  40.7|  13734|  13851|  9987|  2.74|
|  91214|  30356|  42.5|  14642|  15714|  10551|  2.87|
+--------+---------------+---------+----------+-------------+-------
```

We are running Spark here in local mode; however, the true power of Spark comes from its ability to run in the distributed mode and work on large-scale datasets. We will explore distributed features of Spark in the upcoming chapters.

Finding a relationship between data elements

Once we have a decent understanding of the data and some of its main properties, the next step is to find a concrete relationship between data elements. We can use some of the well-established statistical techniques to understand the distribution of data.

Let's continue with our Spark example from the previous section by comparing Total Population to Total Households. We can expect the two numbers to be strongly correlated:

```
println("Covariance: " + df.stat.cov("Total Population", "Total
Households"))
println("Correlation: " + df.stat.corr("Total Population", "Total
Households"))
```

The output from this would be something like this:

```
Covariance: 1.2338126298368526E8
Correlation: 0.9090567549637986
```

As expected, we see the correlation coefficient value closer to 1, indicating a strong correlation between the two variables. The covariance also has a positive value, indicating an increase in one variable would result in the increase of the other variable, and a similar decrease would have the reverse effect.

We can also look at the data in terms of n-tiles. The following code creates 100 tiles ordered by the `Total Population` column:

1. Create a temporary view on top of the Spark DataFrame created in the preceding example. Name this temporary view `tmp_data`:

   ```
   df.createOrReplaceTempView("tmp_data")
   ```

2. Run the Spark SQL on the previously created `tmp_data` view, which uses the window function, `ntile`, which does the following:
   - Orders the data by total population
   - Divides the data into 100 tiles by creating almost equally-sized tiles by starting from the top of the ordered data and going down
   - Selects all columns from the view and additionally computed tile value as the tier

3. Show the contents of the DataFrame output:

   ```
   spark.sql("select *, ntile(100) over(order by `Total Population`)
   tier from tmp_data").show()
   ```

On running the preceding code, the sample output would look something like this:

```
+---------+----------------+----------+------------+-------------+-------
|Zip Code|Total Population|Median Age|Total Males|Total Females|Total
Households|Average Household Size|tier|
+---------+----------------+----------+------------+-------------+-------
|  90079|  0|  0.0|  0|  0|  0|  0.0|  1|
|  90090|  0|  0.0|  0|  0|  0|  0.0|  1|
|  90506|  0|  0.0|  0|  0|  0|  0.0|  1|
|  90747|  0|  0.0|  0|  0|  0|  0.0|  1|
|  90831|  0|  0.0|  0|  0|  0|  0.0|  2|
|  91608|  0|  0.0|  0|  0|  0|  0.0|  2|
|  91371|  1|  73.5|  0|  1|  1|  1.0|  2|
|  90095|  3|  52.5|  2|  1|  2|  1.5|  2|
|  90071|  15|  45.5|  13|  2|  0|  0.0|  3|
|  90822|  117|  63.9|  109|  8|  2|  4.5|  3|
|  91046|  156|  74.0|  51|  105|  114|  1.37|  3|
|  91210|  328|  33.9|  162|  166|  178|  1.84|  3|
|  93563|  388|  44.5|  263|  125|  103|  2.53|  4|
|  91759|  476|  47.2|  239|  237|  216|  2.2|  4|
|  90073|  539|  56.9|  506|  33|  4|  1.25|  4|
|  93544|  1259|  52.4|  689|  570|  569|  2.2|  4|
|  91008|  1391|  54.6|  614|  777|  562|  2.39|  5|
|  90263|  1612|  19.7|  665|  947|  0|  0.0|  5|
|  93243|  1699|  40.9|  884|  815|  623|  2.73|  5|
|  93040|  2031|  29.3|  1052|  979|  522|  3.89|  5|
```

```
+--------+---------------+----------+-----------+-------------+------+
only showing top 20 rows
```

 Note the value of the tier column. It starts from 1 and repeats it for a few rows. It moves on to 2 and repeats it for few rows. At the very end of the output DataFrame, the tier value is going to be 100.

This analysis could be used to cluster zip code by population density. For example, instead of this, we could decide to create only three tiles with low, medium, and high population density. The purpose of the example is to illustrate that Spark provides a comprehensive API set that could be leveraged to establish a relationship between data elements.

Visualizing data

Graphs and charts are used to gain a better understanding of the data relationship. We will use the following to explore data visually:

- Combination of Spark and Vegas viz
- Spark Notebook

Vegas viz for data visualization

Vegas viz is a MatPlotLib-like library for Scala and Spark. The documentation for this library can be found at `https://github.com/vegas-viz/Vegas`. Spark does not contain any built-in support for data visualization. Vegas viz provides a convenient mechanism to add visualization to a Spark program written in Scala.

In order to use this library with Spark, let's add the following dependencies to `build.sbt`:

```
libraryDependencies ++= Seq(
  "org.apache.spark" %% "spark-sql" % "2.4.0", // Spark Core Library
  "org.vegas-viz" %% "vegas-spark" % "0.3.11", // Vegas Viz Library
  "org.scalatest" %% "scalatest" % "3.0.5" % "test" // Scala test library
)
```

Continuing with the example from the previous section, let's say we want to see, visually, the most populated ZIP (90th percentile).

Let's create a Scala program to do so:

1. Import `SparkFiles` from the `spark` package and `SparkSession` from the `spark.sql` package. `SparkFiles` is needed for accessing the CSV file located on the internet. `SparkSession` is needed for creating a Spark session with a program:

   ```
   import org.apache.spark.SparkFiles
   import org.apache.spark.sql.SparkSession
   ```

2. Import Vegas viz packages needed for visualization using Spark:

   ```
   import vegas._
   import vegas.sparkExt._
   ```

3. Create a Scala object for an entry point into the program:

   ```
   object SparkExample {
   ...
   }
   ```

4. Define a method that creates a local Spark session:

   ```
   def getSparkSession(): SparkSession = {
      val spark =
   SparkSession.builder().master("local").getOrCreate()
      spark.sparkContext.setLogLevel("ERROR")
      spark
   }
   ```

5. Define the main method that is the entry point for this program:

   ```
   def main(args: Array[String]): Unit = {
   ...
   }
   ```

6. Implement the main method using the following steps:
   1. Create a Spark session.
   2. Add the internet source file to the Spark context.
   3. Read the contents of the internet source CSV file as a Spark DataFrame, by inferring the first line as the header and inferring the schema from the contents of the CSV.
   4. Create a temporary view on the DataFrame.

5. Run the Spark SQL on the temporary view by using the `ntile` window function.

6. Filter out the tiers that are 90 or above.

7. Use the `Vegas` library's API to create a plot of the filtered DataFrame using the `Zip Code` column as the *x* axis and the `Total Population` column as the *y* axis. The *x* axis data is a discrete number, whereas the *y* axis data is quantity. Mark the plot as a bar chart.

8. Show the plot on a screen.

9. Stop the `SPark` session.

We implement the preceding steps using the following code:

```
    val spark = getSparkSession()
spark.sparkContext.addFile("https://data.lacity.org/api/views/nxs9-385f/row
s.csv")
    val df = spark.read.option("header", true).option("inferSchema",
true).csv(SparkFiles.get("rows.csv"))
    df.createOrReplaceTempView("tmp_data")
    val dfWithTier = spark.sql("select *, ntile(100) over(order by `Total
Population`) tier from tmp_data")
    val dfTier90Plus = dfWithTier.where("tier >= 90")
    val plot = Vegas().withDataFrame(dfTier90Plus).encodeX("Zip Code",
Nom).
      encodeY("Total Population", Quant).
      mark(Bar)
    plot.show
    spark.stop()
```

We can put all of this together and we have a single Spark program that can be executed:

```
import org.apache.spark.SparkFiles
import org.apache.spark.sql.SparkSession
import vegas._
import vegas.sparkExt._

object SparkExample {
  def getSparkSession(): SparkSession = {
    val spark = SparkSession.builder().master("local").getOrCreate()
    spark.sparkContext.setLogLevel("ERROR")
    spark
  }
  def main(args: Array[String]): Unit = {
    val spark = getSparkSession()
spark.sparkContext.addFile("https://data.lacity.org/api/views/nxs9-385f/row
s.csv")
    val df = spark.read.option("header", true).option("inferSchema",
```

```
true).csv(SparkFiles.get("rows.csv"))
    df.createOrReplaceTempView("tmp_data")
    val dfWithTier = spark.sql("select *, ntile(100) over(order by `Total
Population`) tier from tmp_data")
    val dfTier90Plus = dfWithTier.where("tier >= 90")
    val plot = Vegas().withDataFrame(dfTier90Plus).encodeX("Zip Code",
Nom).
        encodeY("Total Population", Quant).
        mark(Bar)
    plot.show
    spark.stop()
  }
}
```

Running the preceding code will produce the following screenshot:

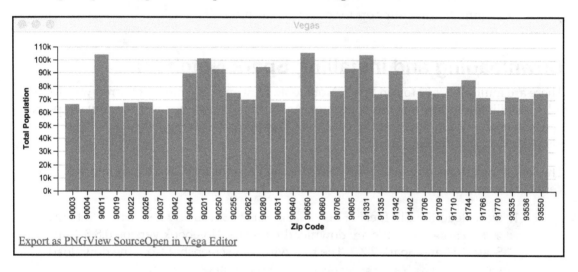

In the aforementioned example, we visually looked at zip codes with the total population in the 90th percentile. Visual methods provide intuition about properties of the data. We can easily conclude that densely populated zip code mostly have similar total populations with an approximate population size of 80 K.

Spark Notebook for data visualization

Spark Notebook (http://spark-notebook.io/) is an open source notebook that provides a web-based interface to perform interactive data analysis. This tool lets a user combine Scala code, SQL queries, Markup, and JavaScript in a collaborative manner to explore, analyze, and learn from data. We will primarily be using this tool for data visualization; however, this can also be used for many other purposes. There are several open source notebook solutions that are available today. However, what is unique are the following features of the tool:

- Scala based
- Excellent integration with Spark
- Support for multiple Spark sessions that are isolated from each other
- Comprehensive support for data visualization

Downloading and installing Spark Notebook

There are multiple ways to run Spark Notebook. However, the most preferred way to run it is by installing it locally on your computer. As of this writing, the most recently supported Spark version is 2.2.0. This is sufficient for us to explore the power of Spark Notebook to visualize data in different forms.

The following are the download and setup instructions for macOS. This should work in a very similar fashion for Linux OS and Windows OS:

1. Download the ZIP file: http://spark-notebook.io/dl/zip/0.8.3/2.11/2.2.2/
   2.7.3/true/true. We are downloading Spark Notebook version 0.8.3, built with
   Scala 2.11 and Spark 2.2.2. The following is a screenshot from the download
   website (http://spark-notebook.io/) that you can refer to:

tgz		deb		zip		docker
Notebook 0.8.3		**Notebook 0.7.0-pre2**		**Notebook 0.8.3**		**Notebook 0.8.3**
Scala 2.11	Scala 2.10	Scala 2.11	Scala 2.10	Scala 2.11	Scala 2.10	Scala 2.11
Spark 2.2.2 and Hadoop	Spark 2.2.2 and Hadoop	Spark 1.6.3 and Hadoop	Spark 1.6.0 and Hadoop	Spark 2.2.2 and Hadoop	Spark 2.2.2 and Hadoop	Spark 2.2.2 and Hadoop
◦ 2.7.2 parquet	◦ 2.7.2 parquet	◦ 2.7.2	◦ 2.6.0 hive parquet	◦ 2.6.0 hive parquet ◦ 2.7.2 hive parquet ◦ 2.7.2 parquet ◦ 2.7.3 hive parquet	◦ 1.0.3 parquet ◦ 2.7.2 parquet	◦ 2.6.0 parquet ◦ 2.6.0-cdh5.10.1 hive parquet ◦ 2.6.0 hive parquet ◦ 2.7.2 parquet

2. Unzip the downloaded ZIP file in a suitable location on your computer:

```
$ unzip spark-notebook-0.8.3-scala-2.11.8-spark-2.2.2-hadoop-2.7.3-
with-hive.zip
```

3. At this point, you should have a directory called `spark-notebook-0.8.3-scala-2.11.8-spark-2.2.2-hadoop-2.7.3-with-hive`. This directory contains the Spark Notebook installation. Change to the binary directory, where the start script is located:

```
$ cd spark-notebook-0.8.3-scala-2.11.8-spark-2.2.2-hadoop-2.7.3-
with-hive/bin
```

4. Start the Spark Notebook server, using the following command:

```
$ bash spark-notebook
```

5. This will start the Spark Notebook server and you will see the output on your Terminal, similar to the following one:

```
Play server process ID is 3744
SLF4J: Class path contains multiple SLF4J bindings.
SLF4J: Found binding in

[jar:file:/Users/rajeshgupta/Downloads/spark-notebook-0.8.3-
scala-2.11.8-spark-2.2.2-hadoop-2.7.3-with-
hive/lib/ch.qos.logback.logback-
classic-1.1.1.jar!/org/slf4j/impl/StaticLoggerBinder.class]

SLF4J: Found binding in
[jar:file:/Users/rajeshgupta/Downloads/spark-notebook-0.8.3-
scala-2.11.8-spark-2.2.2-hadoop-2.7.3-with-
hive/lib/org.slf4j.slf4j-
log4j12-1.7.16.jar!/org/slf4j/impl/StaticLoggerBinder.class]

SLF4J: See http://www.slf4j.org/codes.html#multiple_bindings for an
explanation.
SLF4J: Actual binding is of type
[ch.qos.logback.classic.util.ContextSelectorStaticBinder]
[info] play - Application started (Prod)
[info] play - Listening for HTTP on /0:0:0:0:0:0:0:0:9000
```

The final line indicates that the server can be accessed from the browser on HTTP port 9000. Let this server keep running.

6. Verify that Spark Notebook is running correctly by visiting the landing page from your web browser by going to `http://localhost:900/`. You should see a screenshot similar to the one that follows:

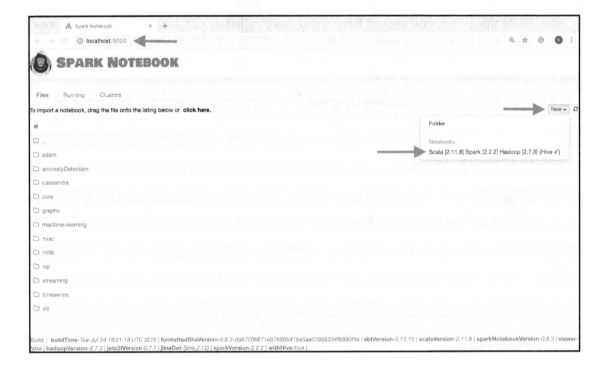

Creating a Spark Notebook with simple visuals

To begin with, we will start by creating a Spark Notebook with some simple visuals:

1. From the landing page shown previously, select **New** | **Scala [2.11.8] Spark [2.2.2] Hadoop [2.7.3] {Hive }** to create a Spark Notebook. Select an appropriate name for the notebook. The following screenshot shows the different prompts:

The output should be similar to the following one on your Terminal:

```
...
[info] application - Creating notebook at /
[info] application - save at path /HandsScalaOnNotebook.snb with message
None
[info] application - listNotebooks at path /
[debug] application - content: /HandsScalaOnNotebook.snb
...
```

2. You should see that `HandsOnScalaNotebook` will appear on the left-hand side of the screen under the **Files** tab. Select this notebook:

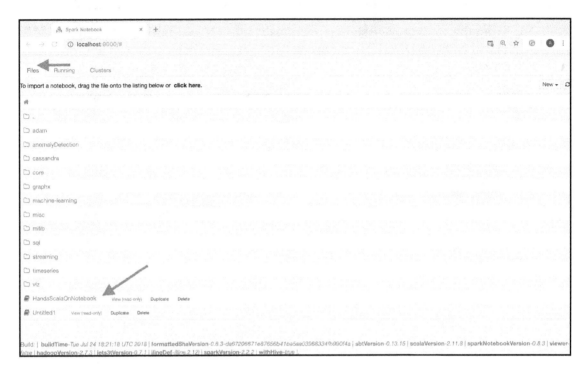

The output should be similar to the following one on your Terminal:

```
. . .
[info] application - getNotebook at path HandsScalaOnNotebook.snb
[info] application - Loading notebook at path HandsScalaOnNotebook.snb
[info] application - Calling action
[info] application - Starting kernel/session because nothing for None and
Some(HandsScalaOnNotebook.snb)
[info] application - Loading notebook at path HandsScalaOnNotebook.snb
. . .
```

3. At this point, your Spark Notebook is running. The following screenshot shows how it would look in the browser with the screen divided into two areas:
   - **Code and Output Area**
   - **Variables and Errors Area**

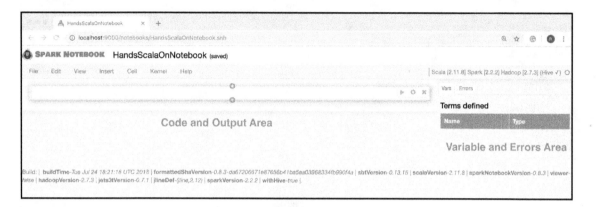

4. Let's create our first plot, which is a simple bar chart, by performing the following steps:
   1. Define a Scala case class as follows in this first cell on the notebook, and then press *SHIFT + ENTER* to execute this cell:

   ```
   case class Person(name: String, age: Int)
   ```

   2. Create a `CustomPlotlyChart` instance with a sequence of `Person` objects as input, the type of plot as a bar chart, the name attribute of `Person` as the *x* axis, and the `age` attribute of `Person` as the *y* axis. Press *SHIFT + ENTER* to execute this cell:

   ```
   CustomPlotlyChart(
     Seq(Person("James Bond", 50), Person("Jon Doe", 25),
   Person("Mickey Mouse", 18), Person("Foo", 33)),
     dataOptions="{type: 'bar'}",
     dataSources="{x: 'name', y: 'age'}")
   ```

5. On the execution of the preceding code, the Spark Notebook's screen should look similar to the following screenshot, with a bar chart plotted:

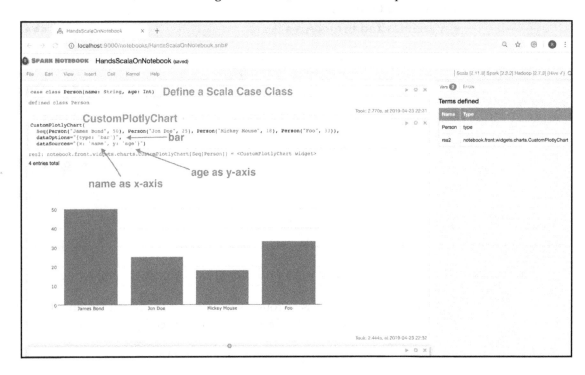

As illustrated previously, using the `CustomPlotlyChart` class of the Spark Notebook library, we are able to create a bar chart with just a single line of code. Here is the complete code:

```scala
case class Person(name: String, age: Int)

CustomPlotlyChart(
  Seq(Person("James Bond", 50), Person("Jon Doe", 25), Person("Mickey
Mouse", 18), Person("Foo", 33)),
  dataOptions="{type: 'bar'}",
  dataSources="{x: 'name', y: 'age'}")
```

In the preceding example, we used a Scala-provided sequence of persons as the data input. We could have instead used a Spark DataFrame as the data input and would have gotten the same chart:

```scala
case class Person(name: String, age: Int)

val persons = Seq(Person("James Bond", 50), Person("Jon Doe", 25),
```

```
Person("Mickey Mouse", 18), Person("Foo", 33)).toDF

CustomPlotlyChart(
    persons,
    dataOptions="{type: 'bar'}",
    dataSources="{x: 'name', y: 'age'}")
```

The following is a screenshot of Spark Notebook with the aforementioned code executed:

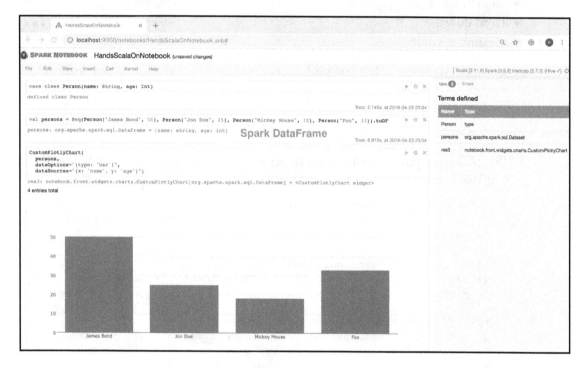

This clearly illustrates that we can use Scala sequences and the Spark DataFrame as data input to the `CustomPlotlyChart` class of the Spark Notebook library. This offers a great deal of flexibility when using this library with data visualization.

Next, we will look at how to create more types of charts using Spark Notebook.

More charts with Spark Notebook

In this section, we will look at more examples of charts using Spark Notebook. The bar chart is certainly very commonly used for data visualization but there are several other types of charts that provide different insights into the data.

Box plot

The box plot is a classic and standardized way of displaying the distribution of data based on the following properties of a dataset:

- Minimum value
- First quartile
- Median value
- Third quartile
- Maximum value

To get a good understanding of the box plot, please refer to `https://www.khanacademy.org/math/statistics-probability/summarizing-quantitative-data/box-whisker-plots/a/box-plot-review`.

Let's explore this with an example:

1. Create three separate Spark DataFrames with 20 random numbers in the range of 0 to 100. Label each creation of the DataFrame as a unique experiment:

```
val df1 = sparkSession.range(20).map(n =>
(scala.util.Random.nextInt(100), "Experiment 1")).toDF("num",
"experiment")

val df2 = sparkSession.range(20).map(n =>
(scala.util.Random.nextInt(100), "Experiment 2")).toDF("num",
"experiment")

val df3 = sparkSession.range(20).map(n =>
(scala.util.Random.nextInt(100), "Experiment 3")).toDF("num",
"experiment")
```

2. Create a combined DataFrame by taking the union of all three DataFrames:

```
val df = df1.union(df2).union(df3)
```

3. Use the box plot to display the properties of each of the three experiments:

```
CustomPlotlyChart(df, dataOptions="{type: 'box', splitBy:
'experiment'}", dataSources="{y: 'num'}")
```

You can run this code in Spark Notebook by executing each of the three steps as an individual cell (using the *SHIFT + ENTER* key combinations) and you will get an output that is similar to the following screenshot:

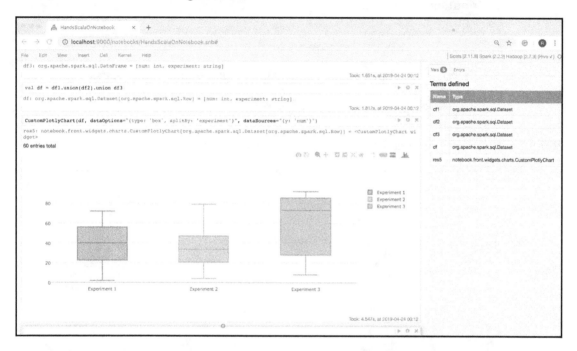

We can see that each of the three experiments has a slightly different distribution of data. This is expected because we are generating pseudorandom numbers with a uniform distribution property. The first experiment produced 20 random number from 0 to 100. The second experiment produced another 20 numbers from 0 to 100. Assuming uniform distribution, the second experiment's numbers would be different from the first one. Similarly, the third experiment's numbers will be different from that of the first and second, again due to the uniform distribution property.

Histogram

A histogram is a representation of the distribution of numerical data. The data is assumed to be continuous and grouped into a certain number of bins. The height of each bin determines the number of occurrences within the established range of the bin.

Let's explore the histogram plot with an example:

1. Create a Spark DataFrame consisting of 1,000 random numbers with Gaussian distribution properties:

```
val df = sparkSession.range(1000).map(n => (n,
scala.util.Random.nextGaussian())).toDF("num", "value")
```

2. Plot histogram of random values with Gaussian distribution properties:

```
CustomPlotlyChart(df, dataOptions="{type: 'histogram', opacity:
0.7}", dataSources="{x: 'value'}")
```

Running the aforementioned code in Spark Notebook results in the following screenshot:

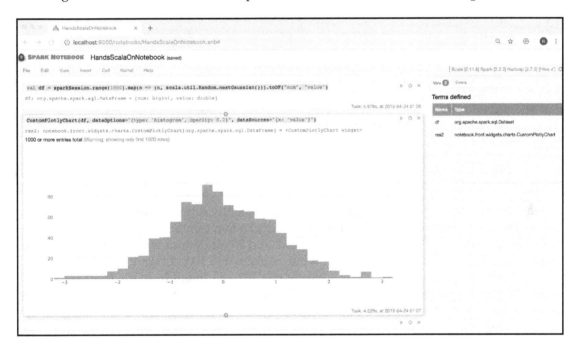

We can easily change the axis from the *x* axis to the *y* axis by doing the following:

```
CustomPlotlyChart(df, dataOptions="{type: 'histogram', opacity: 0.7}",
dataSources="{y: 'value'}")
```

On running this code, the histogram would look like the following screenshot:

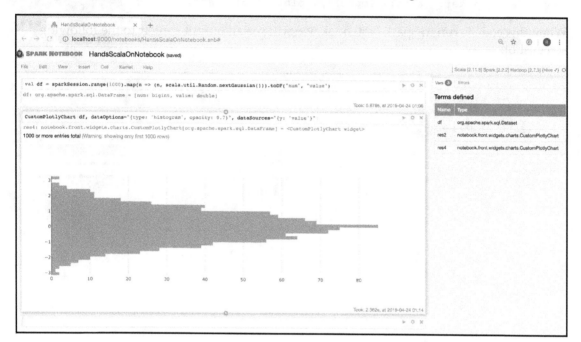

This is the same data, now with the inverted axis. Depending upon the use case, either one can be used. However, the first representation that uses the *x* axis is more common in practice.

Bubble chart

Bubble charts are generally used to represent the impacts of outcomes by representing them as different sized bubbles. The bigger the size of the bubble, the greater the impact.

Let's explore this kind of chart with a simple example:

1. Create a Spark DataFrame consisting of three records with different impacts:

```
val df = Seq((1, 10, 30, "blue"), (2, 11, 60, "green"), (3, 12, 90,
"red")).toDF("x", "y", "impact", "color")
```

2. Create a bubble chart:

```
CustomPlotlyChart(df, layout="{title: 'Impact', showlegend: false,
height: 600, width: 600}",
```

```
dataOptions="{mode: 'markers'}",dataSources="{x: 'x', y: 'y',
marker: {size: 'impact', color: 'color'}}")
```

When this code is run in Spark Notebook, we will see an output similar to the following screenshot:

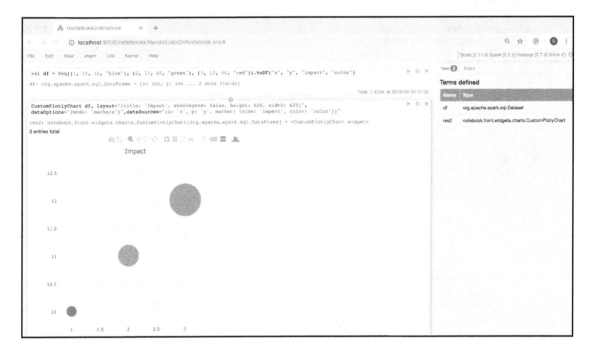

Bubble charts find a significant usage when we use data to communicate a story and want to make it interesting for the target audience who might not be familiar with in-depth details of the underlying data.

Summary

This chapter primarily focused on data exploration and visualization techniques to understand and establish a relationship between data elements. We also learned how to work with samples of data and apply randomness appropriately to select unbiased samples. Understanding properties and the relationship of data are very important because it helps in streamlining the data processing and simplifies the subsequent life cycles of the data.

In the next chapter, we will deep dive into statistical techniques and hypothesis testing.

5
Applying Statistics and Hypothesis Testing

This chapter provides an overview of statistical methods used in data analysis and covers techniques for deriving meaningful insights from data. We will first look at some basic statistical techniques used to gain a better understanding of data before moving on to more advanced methods that are used to compute statistics on vectorized data instead of simple scalar data.

This chapter also covers the various techniques for generating random numbers. Random numbers play a significant part in data analysis because they help us work with sample data in much smaller datasets. A good random sample selection ensures that smaller datasets can act as a good representative of the much bigger dataset.

We will also gain an understanding of hypothesis testing and look at some Scala tools readily available to make this task easier.

The following are the topics that we will be covering in this chapter:

- Basics of statistics
- Vector level statistics
- Random data generation
- Hypothesis testing

Basics of statistics

This section introduces the basics of using applied examples.

Summary level statistics

Summary level statistics provide us with such information as minimum, maximum, and mean values of data.

The following is an example in Spark that looks at summarizing numbers from 1 to 100:

1. Start a Spark shell in your Terminal:

    ```
    $ spark-shell
    ```

2. Import Random from Scala's util package:

    ```
    scala> import scala.util.Random
    import scala.util.Random
    ```

3. Generate integers from 1 to 100 (included) and use the shuffle method of Scala's Random utility class to randomize their positions:

    ```
    scala> val nums = Random.shuffle(1 to 100) // 100 numbers
    randomized
    nums: scala.collection.immutable.IndexedSeq[Int] = Vector(70, 63,
    9, 80, 12, 49, 65, 95, 51, 66, 90, 53, 82, 97, 13, 30, 92, 69, 3,
    7, 11, 72, 37, 16, 48, 75, 100, 88, 78, 38, 91, 35, 26, 56, 58, 47,
    59, 71, 15, 57, 21, 1, 94, 27, 18, 8, 61, 44, 96, 2, 32, 62, 67,
    24, 93, 40, 36, 99, 68, 17, 4, 77, 28, 25, 73, 42, 45, 39, 98, 43,
    20, 41, 34, 33, 86, 84, 64, 31, 87, 29, 10, 81, 55, 52, 76, 89, 23,
    54, 83, 22, 14, 79, 6, 74, 85, 5, 19, 50, 60, 46)
    ```

4. Use Spark's implicit feature to convert the preceding numbers into a Spark DataFrame:

    ```
    scala> val df = nums.toDF()
    df: org.apache.spark.sql.DataFrame = [value: int]
    ```

5. Display the summary from the Spark DataFrame:

    ```
    scala> df.summary().show()
    +-------+-----------------+
    |summary|            value|
    +-------+-----------------+
    |  count|              100|
    ```

```
|  mean|  50.5|
|  stddev|29.011491975882016|
|  min|  1|
|  25%|  25|
|  50%|  50|
|  75%|  75|
|  max|  100|
+-------+------------------+
```

This example is somewhat trivial, as we are working with 100 numbers from 1 to 100. However, it illustrates the power of statistics. In this example, the order of numbers is intentionally randomized to illustrate percentiles, which will be discussed in more detail later.

When we are working with data whose properties are unknown or changing over time, summary level statistics could prove to be a powerful tool to gain insights into data properties in an expeditious manner. Most of these stats can be computed relatively quickly, even on fairly large datasets.

Some additional statistics displayed in the previous example are:

- Mean (average value)
- Standard deviation
- Percentiles (25%, 50%, and 75%)

Computing percentiles is relatively more expensive because it requires data to be sorted.

We can also use Scala collection APIs to get some of this info:

1. Count the numbers:

   ```scala
   scala> val count = nums.size
   count: Int = 100
   ```

2. Compute the mean value of the numbers:

   ```scala
   scala> val mean = nums.sum.toDouble / count
   mean: Double = 50.5
   ```

3. Compute the minimum value of the numbers:

   ```scala
   scala> val min = nums.min
   min: Int = 1
   ```

4. Compute the maximum value of the numbers:

```
scala> val max = nums.max
max: Int = 100
```

5. Sort the numbers in ascending order. This is needed to compute percentiles in the next steps:

```
scala> val numsSorted = nums.sorted // sorting is needed for
computing percentiles
numsSorted: scala.collection.immutable.IndexedSeq[Int] = Vector(1,
2, 3, 4, 5, 6, 7, 8, 9, 10, 11, 12, 13, 14, 15, 16, 17, 18, 19, 20,
21, 22, 23, 24, 25, 26, 27, 28, 29, 30, 31, 32, 33, 34, 35, 36, 37,
38, 39, 40, 41, 42, 43, 44, 45, 46, 47, 48, 49, 50, 51, 52, 53, 54,
55, 56, 57, 58, 59, 60, 61, 62, 63, 64, 65, 66, 67, 68, 69, 70, 71,
72, 73, 74, 75, 76, 77, 78, 79, 80, 81, 82, 83, 84, 85, 86, 87, 88,
89, 90, 91, 92, 93, 94, 95, 96, 97, 98, 99, 100)
```

6. Compute the 25 percentile:

```
scala> val percent25 = numsSorted(count/4-1)
percent25: Int = 25
```

7. Compute the 50 percentile:

```
scala> val percent50 = numsSorted(count/2-1)
percent50: Int = 50
```

8. Finally, compute the 75 percentile:

```
scala> val percent75 = numsSorted((3*count/4)-1)
percent75: Int = 75
```

Let's look at how percentiles are calculated in the preceding example. The first step is to sort the data. Once the data is sorted, we can visualize the percentiles, as shown in the following screenshot:

```
numsSorted: scala.collection.immutable.IndexedSeq[Int] = Vector(1, 2, 3, 4, 5, 6, 7, 8, 9, 10,
11, 12, 13, 14, 15, 16, 17, 18, 19, 20, 21, 22, 23, 24, 25, 26, 27, 28, 29, 30, 31, 32, 33, 34,
35, 36, 37, 38, 39, 40, 41, 42, 43, 44, 45, 46, 47, 48, 49, 50, 51, 52, 53, 54, 55, 56, 57, 58,
59, 60, 61, 62, 63, 64, 65, 66, 67, 68, 69, 70, 71, 72, 73, 74, 75, 76, 77, 78, 79, 80, 81, 82,
83, 84, 85, 86, 87, 88, 89, 90, 91, 92, 93, 94, 95, 96, 97, 98, 99, 100)
```

- The value at the 25% mark is the 25 percentile
- At the 50% mark, it is the 50 percentile
- At the 75% mark, it is the 75 percentile

The value at the 50% mark is also known as the **median value**.

Percentiles help us in understanding data skews to a fairly good degree. Sorting is generally a fairly expensive operation with most sort algorithms requiring *O(n log n)* compute times. This is why computing exact median and percentile values for fairly large datasets is generally a slow process. For many practical purposes, an approximate mean is generally sufficient and there are many frameworks that are able to compute this value efficiently and quickly.

We can certainly compute other statistical properties, such as standard deviation using plain Scala code. However, Scala frameworks, such as Saddle, Breeze, Scalalab, and Spark, have built-in support to easily compute values. With a framework such as Spark, the true power comes from computing these values at scale in a distributed computing environment.

Let's look at another way to gather statistics using Spark's built-in Statistics API:

1. Try the following in your `spark-shell`:

   ```
   $ spark-shell
   ```

2. Import `Vectors` from Spark MLLib's `linalg` package:

   ```
   scala> import org.apache.spark.mllib.linalg.Vectors
   import org.apache.spark.mllib.linalg.Vectors
   ```

3. Import the `Statistics` class from Spark MLLib's `stat` package:

   ```
   scala> import org.apache.spark.mllib.stat.Statistics
   import org.apache.spark.mllib.stat.Statistics
   ```

4. Create a Spark RDD of vectors of size five RDD:

   ```
   scala> val data = sc.parallelize(
    | Seq(
    | Vectors.dense(1.0, 0.0, 0.0, 1.0, 100.0),
    | Vectors.dense(2.0, 1.0, 10.0, 10.0, 200.0),
    | Vectors.dense(3.0, 2.0, 20.0, 100.0, 300.0),
    | Vectors.dense(4.0, 3.0, 30.0, 1000.0, 400.0),
    | Vectors.dense(5.0, 4.0, 40.0, 10000.0, 500.0),
    | Vectors.dense(6.0, 5.0, 50.0, 100000.0, 600.0),
    | Vectors.dense(7.0, 6.0, 60.0, 1000000.0, 700.0),
    | Vectors.dense(8.0, 7.0, 70.0, 10000000.0, 800.0),
    | Vectors.dense(9.0, 8.0, 80.0, 100000000.0, 900.0),
    | Vectors.dense(9.9, 9.0, 90.0, 1000000000.0, 1000.0)
    | )
    | )
   ```

```
data:
org.apache.spark.rdd.RDD[org.apache.spark.mllib.linalg.Vector] =
ParallelCollectionRDD[0] at parallelize at <console>:26
```

5. Compute the column summary statistics on Spark RDD, that we created previously:

```
// Compute column summary statistics
scala> val summary = Statistics.colStats(data)
summary: org.apache.spark.mllib.stat.MultivariateStatisticalSummary
= org.apache.spark.mllib.stat.MultivariateOnlineSummarizer@7b13ae73
```

6. Get the number of records:

```
scala> summary.count // number of records
res0: Long = 10
```

7. Get the mean value for each column:

```
scala> summary.mean // mean value for each column
res1: org.apache.spark.mllib.linalg.Vector =
[5.489999999999999,4.5,45.0,1.111111111E8,550.0]
```

8. Get the column-wise minimum:

```
scala> summary.min // column-wise min
res2: org.apache.spark.mllib.linalg.Vector =
[1.0,0.0,0.0,1.0,100.0]
```

9. Get the column-wise maximum:

```
scala> summary.max // column-wise max
res3: org.apache.spark.mllib.linalg.Vector =
[9.9,9.0,90.0,1.0E9,1000.0]
```

10. Get the column-wise norm L1:

```
scala> summary.normL1 // column-wise norm L1
res4: org.apache.spark.mllib.linalg.Vector =
[54.9,45.0,450.0,1.111111111E9,5500.0]
```

11. Get the column-wise L2, aka the Euclidean norm :

```
scala> summary.normL2 // column-wise Euclidean magnitude
res5: org.apache.spark.mllib.linalg.Vector =
[19.570641277178424,16.881943016134134,168.81943016134133,1.0050378
15259212E9,1962.1416870348583]
```

12. Get the column-wise variance:

```
scala> summary.variance // column-wise variance
res6: org.apache.spark.mllib.linalg.Vector =
[9.067666666666666,9.166666666666666,916.6666666666666,9.8516024444
69384E16,91666.66666666667]
```

13. Get the column-wise count of the number of non-zeros:

```
scala> summary.numNonzeros // column-wise count of non-zero values
res7: org.apache.spark.mllib.linalg.Vector =
[10.0,9.0,9.0,10.0,10.0]
```

In the preceding example, we created an RDD consisting of 10 records and five columns with all numerical values. As can be seen, the `Statistics` API of Spark is another convenient way to compute statistics. The `Statistics` API has a few other useful functionalities, but we will look at the correlation computation API in particular.

Correlation statistics

In the previous section, we have been looking at each column or variable from the RDD in isolation. Statistical values, such as mean, median, min, or max, for each column are completely independent of any other column in the dataset. In this section, we will look at a correlation where we want to measure how strongly two columns or variables are related to each other.

We will be using Spark's `Statistics` API to compute the correlation. Try the following in `spark-shell`:

1. Start a Spark shell in your Terminal:

   ```
   $ spark-shell
   ...
   ```

2. Import the `Vectors` class from Spark MLLib's `linalg` package:

   ```
   scala> import org.apache.spark.mllib.linalg.Vectors
   import org.apache.spark.mllib.linalg.Vectors
   ```

3. Import the `Statistics` class from Spark MLLib's `stat` package:

   ```
   scala> import org.apache.spark.mllib.stat.Statistics
   import org.apache.spark.mllib.stat.Statistics
   ```

4. Create a Spark RDD of vectors with three elements each:

```
scala> val data = sc.parallelize(
     | Seq(
     | Vectors.dense(0.0, 1.0, 100.0),
     | Vectors.dense(10.0, 10.0, 200.0),
     | Vectors.dense(20.0, 100.0, 300.0),
     | Vectors.dense(30.0, 1000.0, 400.0),
     | Vectors.dense(40.0, 10000.0, 500.0),
     | Vectors.dense(50.0, 100000.0, 600.0),
     | Vectors.dense(60.0, 1000000.0, 700.0),
     | Vectors.dense(70.0, 10000000.0, 800.0),
     | Vectors.dense(80.0, 100000000.0, 900.0),
     | Vectors.dense(90.0, 1000000000.0, 1000.0)
     | )
     | )
 data:
org.apache.spark.rdd.RDD[org.apache.spark.mllib.linalg.Vector] =
ParallelCollectionRDD[0] at parallelize at <console>:26
```

5. Compute the pair-wise column correlation using the Statistics .corr method:

```
scala> val correlation = Statistics.corr(data) // Compute
correlation
correlation: org.apache.spark.mllib.linalg.Matrix =
1.0 0.5701779377812776 1.0
0.5701779377812776 1.0 0.5701779377812777
1.0 0.5701779377812777 1.0
```

In the preceding example, we have a dataset consisting of 10 rows and three columns. We are trying to compute the correlation between these three columns in a pair-wise fashion. We can visually see that column one and column three have a strong correlation, whereas column two does not strongly correlate to either column one or column three.

In the preceding example, we used Pearson's correlation method. This method results in a value between +1 and -1. A value closer to +1 indicates a strong positive correlation between the pair, closer to 0 indicates no correlation, and closer to -1 indicates a strong negative correlation. In fact, we use the following simple linear relationship between column one and column three that has Pearson's correlation values of 1:

```
y =  10x + 100
```

In this case, x represents column one and y represents column three.

Now let's look at a negative correlation example by changing column one values to all negative values:

1. Start a Spark shell in your Terminal:

   ```
   $ spark-shell
   ```

2. Import the Vectors class from Spark MLLib's linalg package:

   ```
   scala> import org.apache.spark.mllib.linalg.Vectors
   import org.apache.spark.mllib.linalg.Vectors
   ```

3. Import the Statistics class from Spark MLLib's stat package:

   ```
   scala> import org.apache.spark.mllib.stat.Statistics
   import org.apache.spark.mllib.stat.Statistics
   ```

4. Create a Spark RDD consisting of vectors with three elements each:

   ```
   scala> val data = spark.sparkContext.parallelize(
   | Seq(
   | Vectors.dense(0.0, 1.0, 100.0),
   | Vectors.dense(-10.0, 10.0, 200.0),
   | Vectors.dense(-20.0, 100.0, 300.0),
   | Vectors.dense(-30.0, 1000.0, 400.0),
   | Vectors.dense(-40.0, 10000.0, 500.0),
   | Vectors.dense(-50.0, 100000.0, 600.0),
   | Vectors.dense(-60.0, 1000000.0, 700.0),
   | Vectors.dense(-70.0, 10000000.0, 800.0),
   | Vectors.dense(-80.0, 100000000.0, 900.0),
   | Vectors.dense(-90.0, 1000000000.0, 1000.0)
   | )
   | )
   data:
   org.apache.spark.rdd.RDD[org.apache.spark.mllib.linalg.Vector] =
   ParallelCollectionRDD[0] at parallelize at <console>:25
   ```

5. Compute a pair-wise column correlation using the Statistics .corr method:

   ```
   scala> Statistics.corr(data)
   res0: org.apache.spark.mllib.linalg.Matrix =
   1.0 -0.5701779377812776 -1.0
   -0.5701779377812776 1.0 0.5701779377812777
   -1.0 0.5701779377812777 1.0
   ```

As can be seen, the correlation value between column one and column three has changed to -1 and this can be represented using the following linear relationship:

```
y = -10x + 100
```

Correlation statistics provide a powerful mechanism with which to observe how strongly two variables are related to each other.

Vector level statistics

In the previous section, we looked at statistics for columns containing a single numeric value. It is often the case that, for **machine learning (ML)**, a more common way to represent data is as vectors of multiple numeric values. A vector is a generalized structure that consists of one or more elements of the same data type. For example, the following is an example of a vector of three elements of type double:

```
[2.0,3.0,5.0]
[4.0,6.0,7.0]
```

Computing statistics in the classic way won't work for vectors. It is also quite common to have weights associated with these vectors. There are times when the weights have to considered as well while computing statistics on such a data type.

Spark MLLib's `Summarizer` (`https://spark.apache.org/docs/latest/api/java/org/apache/spark/ml/stat/Summarizer.html`) provides several convenient methods to compute stats on vector-based data. Let's explore this in a Spark shell:

1. Import the necessary classes from the Spark `ml` package:

```
scala> import org.apache.spark.ml.linalg.{Vector, Vectors}
import org.apache.spark.ml.linalg.{Vector, Vectors}
scala> import org.apache.spark.ml.stat.Summarizer
import org.apache.spark.ml.stat.Summarizer
```

2. Create a Spark DataFrame, consisting of three feature vectors and the weight:

```
scala> val df = Seq(
|    (Vectors.dense(1.0, 2.0, 3.0), 9.0),
|    (Vectors.dense(4.0, 5.0, 6.0), 5.0),
|    (Vectors.dense(7.0, 8.0, 9.0), 1.0),
|    (Vectors.dense(0.0, 1.0, 2.0), 7.0)
| ).toDF("features", "weight")
df: org.apache.spark.sql.DataFrame = [features: vector, weight:
double]
```

3. Display the contents of the Spark DataFrame:

```
scala> df.show(truncate=false)
+-------------+------+
|features     |weight|
+-------------+------+
|[1.0,2.0,3.0]|9.0   |
|[4.0,5.0,6.0]|5.0   |
|[7.0,8.0,9.0]|1.0   |
|[0.0,1.0,2.0]|7.0   |
+-------------+------+
```

4. Use `SummaryBuilder` (https://spark.apache.org/docs/latest/api/java/ org/apache/spark/ml/stat/Summarizer.html) to build min, max, mean, and variance metrics, using `features` and `weight` columns. Summarizer's metrics return a `SummaryBuilder` object that provides summary statistics about a given column:

```
scala> val summarizer = Summarizer.metrics("min", "max", "mean",
                        "variance").summary($"features", $"weight")
summarizer: org.apache.spark.sql.Column =
aggregate_metrics(features, weight)
```

5. Apply `summarizer` on the source DataFrame to create a Summary DataFrame:

```
scala> val summaryDF = df.select(summarizer.as("summary"))
summaryDF: org.apache.spark.sql.DataFrame = [summary: struct<min:
vector, max: vector ... 2 more fields>]
```

6. Display the contents of the Summary DataFrame:

```
scala> summaryDF.show(truncate=false)
+----------------------------------------------------------------
+
|summary
|
+----------------------------------------------------------------
+
|[[0.0,1.0,2.0], [7.0,8.0,9.0],
[1.6363636363636362,2.6363636363636362,3.636363636363636],
[5.304878048780488,5.304878048780488,5.304878048780487]]|
+----------------------------------------------------------------
+
```

7. Extract the `mean` and `variance` with weight from the Summary DataFrame:

```scala
scala> val (min1, max1, meanWithWeight1, varianceWithWeight1) =
summaryDF.select("summary.min", "summary.max", "summary.mean",
"summary.variance").as[(Vector, Vector, Vector, Vector)].first()
min1: org.apache.spark.ml.linalg.Vector = [0.0,1.0,2.0]
max1: org.apache.spark.ml.linalg.Vector = [7.0,8.0,9.0]
meanWithWeight1: org.apache.spark.ml.linalg.Vector =
[1.6363636363636362,2.6363636363636362,3.636363636363636]
varianceWithWeight1: org.apache.spark.ml.linalg.Vector =
[5.304878048780488,5.304878048780488,5.304878048780487]
```

8. Compute the mean and variance with weight using another approach:

```scala
scala> val (min2, max2, meanWithWeight2, varianceWithWeight2) =
df.select(Summarizer.min($"features"), Summarizer.max($"features"),
Summarizer.mean($"features", $"weight"),
Summarizer.variance($"features", $"weight")).as[(Vector, Vector,
Vector, Vector)].first()
min2: org.apache.spark.ml.linalg.Vector = [0.0,1.0,2.0]
max2: org.apache.spark.ml.linalg.Vector = [7.0,8.0,9.0]
meanWithWeight2: org.apache.spark.ml.linalg.Vector =
[1.6363636363636362,2.6363636363636362,3.636363636363636]
varianceWithWeight2: org.apache.spark.ml.linalg.Vector =
[5.304878048780488,5.304878048780488,5.304878048780487]
```

9. Compute a simple mean and variance without weight:

```scala
scala> val (min, max, mean, variance) =
df.select(Summarizer.min($"features"), Summarizer.max($"features"),
Summarizer.mean($"features"),
Summarizer.variance($"features")).as[(Vector, Vector, Vector,
Vector)].first()
min: org.apache.spark.ml.linalg.Vector = [0.0,1.0,2.0]
max: org.apache.spark.ml.linalg.Vector = [7.0,8.0,9.0]
mean: org.apache.spark.ml.linalg.Vector = [3.0,4.0,5.0]
variance: org.apache.spark.ml.linalg.Vector = [10.0,10.0,10.0]
```

In the preceding example, we computed the `min`, `max`, `mean`, and `variance` of vectorized data with and without weight. As can be seen, the results are quite different for mean and variance values when weights are used.

Spark MLLib's `Summarizer` tool (`https://spark.apache.org/docs/latest/api/java/org/apache/spark/ml/stat/Summarizer.html`) provides both variants of the statistical function, one without weights and the other one with weights:

Function name	Computation
count	The count of all vectors seen
max	A vector that contains the maximum for each coefficient
mean	A vector that contains the coefficient-wise mean
min	A vector that contains the minimum for each coefficient
normL1	The L1 norm of each coefficient (sum of the absolute values)
normL2	The L2, aka Euclidean norm, for each coefficient
numNonZeros	A vector with the number of non-zeros for each coefficient
variance	A vector that contains the coefficient-wise variance

The metrics method is more generalized and accepts one or more of the following arguments as a string (case-sensitive):

- `count`: The count of all vectors seen
- `max`: The maximum for each coefficient
- `mean`: A vector that contains the coefficient-wise mean
- `min`: The minimum for each coefficient
- `normL1`: The L1 norm of each coefficient (sum of the absolute values)
- `normL2`: L2, aka the Euclidean norm, for each coefficient
- `numNonzeros`: A vector with the number of non-zeros for each coefficient
- `variance`: A vector that contains the coefficient-wise variance

`Summarizer` tools offer a choice between metrics and individual methods. Either can be used based on personal preference.

Random data generation

Random data generation is useful for several purposes and plays a significant role in performance testing. This technique is also useful for generating synthetic data that can be used for various simulation experiment purposes. In fact, it is randomness that facilitates an unbiased sample selection from a large dataset.

We will look at random data generation with some specific properties:

- Pseudorandom with no specific distribution
- Normal distribution
- Poisson distribution

Pseudorandom numbers

Scala provides built-in support to generate pseudorandom numbers using the `scala.util.Random` class. Let's explore some features of this class using Scala REPL:

1. Import the `Random` class from the `scala.util` package:

```
scala> import scala.util.Random
import scala.util.Random
```

2. Generate 10 random numbers:

```
scala> Range(0, 10).map(i => Random.nextDouble())
res0: scala.collection.immutable.IndexedSeq[Double] =
Vector(0.5538242600417229, 0.5267086862614716, 0.4812270209045445,
0.008044846025885621, 0.48136489192085685, 0.1714965492674987,
0.9854714710135378, 0.2758151704280012, 0.23205567318485132,
0.42791101504509277)
```

In the preceding example, we generated 10 random double values using the `nextDouble` API, which provides uniformly distributed pseudorandom numbers between 0 and 1.

The `scala.util.Random` class has several methods to provides random values of different types:

1. Generate a random integer:

```
scala> Random.nextInt() // Random Integer
res1: Int = -116922537
```

2. Generate a random number with a Gaussian distribution property:

```
scala> Random.nextGaussian() // Random Gaussian (normally)
distributed with a mean of 0.0 and standard deviation of 1.0
res2: Double = -1.1399390663617412
```

3. Generate a random Boolean value:

```
scala> Random.nextBoolean() // Random true or false
res3: Boolean = true
```

4. Generate a random printable character:

```
scala> Random.nextPrintableChar() // Random Char
res4: Char = i
```

5. Generate a random string with 10 characters:

```
scala> Random.nextString(10) // Random String of size 10
res5: String = ▯鸛曜뫡唫勢▯꽈▯堺
```

As can be seen, the Scala `Random` API is quite rich and handy for working with random numbers. If you already have some data for which you want to randomize the order, there is a shuffle API you can use to perform precisely that operation. Let's explore some features of `shuffle` using Scala REPL:

1. Randomize the position of numbers 0 to 9:

```
scala> Random.shuffle(Range(0,10).toList) // Randomize numbers 0 to
9
res7: List[Int] = List(2, 5, 4, 9, 8, 6, 3, 1, 7, 0)
```

2. Randomize the positions of elements in a list of strings:

```
// Randomize list of currencies
scala> Random.shuffle(List("USD", "INR", "EUR", "DKR", "CAD",
"AUD"))
res9: List[String] = List(AUD, INR, EUR, DKR, CAD, USD)
```

3. Randomize the position of elements in a list consisting of mixed types:

```
// Randomize a list of mixed data types
scala> Random.shuffle(List(0, "Jon", 2, "Doe"))
res10: List[Any] = List(2, Jon, 0, Doe)
```

The `shuffle` method provides a convenient way of randomizing the order of elements in an existing collection.

Random numbers with normal distribution

In a Gaussian or normal distribution, the data follows a bell-shaped curve. A normally distributed random is quite useful in statistical analysis because a lot of real-world data exhibits this property.

Let's first re-explore `scala.util.Random` to generate a series of normally distributed random numbers:

1. Import the `Random` class from the `scala.util` package:

```scala
scala> import scala.util.Random
import scala.util.Random
```

2. Generate 20 random numbers with a Gaussian distribution property:

```scala
scala> val num20 = Range(0, 20).map(i => Random.nextGaussian()) //
Random Gaussian (normally) distributed with a mean of 0.0 and
standard deviation of 1.0
num20: scala.collection.immutable.IndexedSeq[Double] =
Vector(0.20633575917837435, 0.7856945092974417, 1.2432023260888005,
-0.26028288029552493, -1.1672076588249276, -1.1057648961382314,
-0.0024377048350471293, 0.18768703356333027, -0.25773643701036303,
0.6731493330185354, -0.5045811414092171, -1.9258127364625324,
-1.5583448873537717, 0.13111785798881095, 0.16927327531581107,
0.6311168720217485, 0.3120733937326494, -1.0091494950203739,
1.500548733195163, 0.5424493305037464)
```

3. Compute the mean of the random numbers generated previously:

```scala
scala> val mean = num20.sum/num20.size // Recompute mean
mean: Double = -0.07043347067227887
```

In the preceding example, we created a collection of 20 random numbers with Gaussian distribution properties and recomputed the mean value. The mean value is closer to 0.0; however, it is not exactly 0.

Let's explore another way to generate normally distributed random numbers. This time, we will be using Spark for this purpose. The following is an example of generating random numbers with normal distribution in Spark:

1. Start a Spark shell in a Terminal:

```
$ spark-shell
```

2. Import `RandomRDD` from Spark MLLib's `random` package:

```
scala> import org.apache.spark.mllib.random.RandomRDDs._
import org.apache.spark.mllib.random.RandomRDDs._
```

3. Create an RDD of 20 random numbers using `normalRDD` from the `RandomRDD` class imported previously:

```
scala> val norm20 = normalRDD(sc, 20)
norm20: org.apache.spark.rdd.RDD[Double] = RandomRDD[0] at RDD at
RandomRDD.scala:42
```

4. Collect the RDD data into a Spark driver:

```
scala> val data = norm20.collect
data: Array[Double] = Array(0.9002432879145585, 1.2017829268140054,
0.22138020379583687, 0.162056540782974, 0.08797635729660246,
0.7485504161725681, -1.4444317942193088, -1.2053105014796643,
1.4366170997899934, -1.1217899878597575, -0.5402419965639337,
-0.39353597754494823, 1.2389234612716393, 0.48195007284822516,
-1.5520071929920085, -0.8154961830848371, 0.7595546221214476,
-0.6509633621290518, -0.020977133758767988, 0.4244958376997622)
```

5. Compute the `mean` value using plain Scala code (without using Spark):

```
scala> val mean = data.sum/data.size // Recompute mean
mean: Double = -0.004061165156233221
```

6. Use RDD's `stats` method to compute statistics, such as `count`, `mean`, `stdev`, `min`, and `max`:

```
scala> norm20.stats
res2: org.apache.spark.util.StatCounter = (count: 20, mean:
-0.004061, stdev: 0.900782, max: 1.436617, min: -1.552007)
```

In the preceding example, we created a normally distributed collection with 20 random values using Spark. We recomputed the mean and observed that its value is a lot closer to 0.0 compared to the earlier method. The biggest benefit of using a framework such as Spark is that it provides higher-level APIs that are very suitable for data analysis. Yet another benefit of Spark is its support for distributed computing at a large scale.

Random numbers with Poisson distribution

Poisson distributions are useful for modeling the occurrence frequency of events within time intervals. Similar to normally distributed data, Poisson distributed data is found quite a lot in real-world scenarios.

The following is an example of generating random numbers with Poisson distribution in Spark:

1. Start a Spark shell in a Terminal:

   ```
   $ spark-shell
   ```

2. Import `RandomRDD` from Spark's MLLib's `random` package:

   ```
   scala> import org.apache.spark.mllib.random.RandomRDDs._
   import org.apache.spark.mllib.random.RandomRDDs._
   ```

3. Set the desired mean value to `1.0`:

   ```
   scala> val mean = 1.0 // desired mean
   mean: Double = 1.0
   ```

4. Create an RDD of 20 random numbers with a Poisson distribution property:

   ```
   scala> val poi20 = poissonRDD(sc, mean, 20) // 20 values
   poi20: org.apache.spark.rdd.RDD[Double] = RandomRDD[0] at RDD at
   RandomRDD.scala:42
   ```

5. Insert the RDD created previously into the Spark driver:

   ```
   scala> poi20.collect
   res0: Array[Double] = Array(0.0, 0.0, 1.0, 0.0, 3.0, 0.0, 0.0, 0.0,
   0.0, 2.0, 1.0, 0.0, 3.0, 1.0, 1.0, 0.0, 3.0, 1.0, 2.0, 0.0)
   ```

6. Compute the mean of the numbers using regular Scala code:

   ```
   scala> val actualMean = data.sum/data.size
   actualMean: Double = 0.9
   ```

7. Compute RDD stats:

   ```
   scala> poi20.stats
   res1: org.apache.spark.util.StatCounter = (count: 20, mean:
   0.900000, stdev: 1.090871, max: 3.000000, min: 0.000000)
   ```

In the preceding example, we are easily able to create a random dataset with Poisson distribution properties. The actual mean value of 0.9 is close to the desired mean value of 1.0. The advantage of using a framework such as Spark is that it provides a higher-level abstraction that eases random data generation with such properties and, because of its distributed nature, Spark also provides the scale necessary to deal with large datasets.

Hypothesis testing

Hypothesis testing is a statistical tool that is used for the following purposes:

- Determining whether a result or model is statistically significant or not
- Ensuring that a result or model did not occur by chance

A statistical hypothesis is used to establish a relationship between data using a sample set of observations. We can call this relationship a result or a model. The goal of hypothesis testing is to eliminate cases where a result occurs by chance. A null hypothesis, on the other hand, establishes that the relationship is not statistically significant.

We typically start with a sample set of observations that consists of values associated with more than one variable. In the *Basics of statistics* section, we looked at properties of a single variable in isolation, except for Pearson's correlation methodology, where we measured the linear relationship between two variables. The reliability of this relationship is, however, dependent on the number of observations used to derive this value. We perform a hypothesis test to confirm the significance of this correlation to determine whether the relationship in the sample data is representative of the relationship in the population data.

Let's look at an example in Spark using Pearson's chi-squared tests for goodness of fit. We will be using Pearson's method:

1. Start a Spark shell in a Terminal:

   ```
   $ spark-shell
   ```

2. Import the necessary classes from Spark's MLLib package:

   ```scala
   scala> import org.apache.spark.mllib.linalg._
   import org.apache.spark.mllib.linalg._

   scala> import org.apache.spark.mllib.regression.LabeledPoint
   import org.apache.spark.mllib.regression.LabeledPoint

   scala> import org.apache.spark.mllib.stat.Statistics
   import org.apache.spark.mllib.stat.Statistics
   ```

```
scala> import org.apache.spark.mllib.stat.test.ChiSqTestResult
import org.apache.spark.mllib.stat.test.ChiSqTestResult

scala> import org.apache.spark.rdd.RDD
import org.apache.spark.rdd.RDD
```

3. Create a sample observation of vectors:

```
scala> val observations = Vectors.dense(0.9, 0.8, 0.7, 0.6, 0.5,
0.4, 0.3, 0.2, 0.1)
observations: org.apache.spark.mllib.linalg.Vector =
[0.9,0.8,0.7,0.6,0.5,0.4,0.3,0.2,0.1]
```

4. Run the chi-square test on the data:

```
scala> Statistics.chiSqTest(observations)
res0: org.apache.spark.mllib.stat.test.ChiSqTestResult =
Chi squared test summary:
method: pearson
degrees of freedom = 8
statistic = 1.2000000000000002
pValue = 0.996641931146752
No presumption against null hypothesis: observed follows the same
distribution as expected..
```

Summary

Statistics play an important role in the data analysis life cycle. This chapter provided an overview of basic statistics. We also learned how to extend basic statistical techniques and use them on data that is represented as vectors. In the vector bases stats, we got some insights into how weights could significantly alter statistical outcomes. We also learned various techniques for random data generation, and, finally, we took a high-level view of how to perform hypothesis testing.

In the next chapter, we will focus on Spark, a distributed data analysis and processing framework.

Section 2: Advanced Data Analysis and Machine Learning

2

In this section, you will do data analysis on distributed data, and get introduced to Spark, a Scala-based distributed framework. This section will cover some interesting **machine learning** (**ML**) concepts such as decision trees, random forests, lasso regression, and k-means clustering.

This section will contain the following chapters:

- Chapter 6, *Introduction to Spark for Distributed Data Analysis*
- Chapter 7, *Traditional Machine Learning for Data Analysis*

6
Introduction to Spark for Distributed Data Analysis

In the previous chapters, we looked at various aspects of the data analysis life cycle using Scala and some of the associated Scala libraries for data analysis. These libraries work well on a single machine; however, most of the real-world data is generally too big to fit into a single machine and requires distributed data processing on multiple machines. It is certainly possible to write distributed data processing code using Scala, but the complexity of handling failures rises significantly in a distributed environment. Fortunately, there are some robust and proven open source solutions that are available to facilitate distributed data processing on a large scale. One such open source solution is Apache Spark.

Apache Spark (https://spark.apache.org/) is a unified analytics engine that supports robust and reliable distributed data processing. It is certainly possible to use Spark on a single machine, and we have already used Spark this way for some of our examples in previous chapters. Using Spark this way is considered a local mode of operation and has its own usefulness in many ways, as we will see later in the chapter. In this chapter, we will explore some features of Spark that make it a truly powerful platform for large-scale distributed data processing. This chapter will provide the transition to performing data analysis on distributed systems and doing it at scale. It will provide a good introduction to Spark, a Scala-based distributed framework for data processing. It will guide the user through setting up Spark on their computer and introduce key Spark features using practical examples.

Apache Spark (https://spark.apache.org/) aims to be a unified analytics engine for large-scale data processing. As we know, there are different kinds of tasks that need to be performed in the data life cycle. One option is to have a separate solution that specializes in specific tasks. Apache Spark takes a different approach by providing a unified engine that has support for the most important data life cycle tasks—for example, data exploration can be easily performed on Spark by pulling data from various sources. At the same time, **machine learning (ML)** tasks can also be performed on the same engine with ease.

Data pipelines can also be run on Spark for performing a simple task, such as data extraction to more sophisticated stream-oriented processing. The biggest benefit of this unified engine is that multiple stakeholders of the data life cycle use a single platform, which helps in the acceleration of data-oriented solutions development and deployment. Another great benefit of Spark is its API support for multiple programming languages. This certainly helps in lowering the barrier for the adoption of this technology in an enterprise. It has support for the following languages:

- **Scala and Java**: **Java Virtual Machine** (**JVM**) languages that are an excellent choice for building robust data pipelines
- **Python**: Excellent for ad hoc analysis
- **R**: An alternative to Python for ad hoc analysis

Spark provides APIs for batch as well as stream data processing in a distributed computing environment. Spark's APIs could be broadly divided into the following five categories:

- **Core**: **Resilient distributed datasets** (**RDD**)
- **SQL**: DataFrames, dataset API
- **Streaming**: Structured Streaming and **Discretized Stream** (**DStream**)
- **MLlib**: ML
- **GraphX**: Graph processing

Apache Spark is a very active open source project. New features and performance improvements are made on a regular basis. Typically, there is a new release of Apache Spark every three months. At the time of this writing, 2.4.0 is the most recent release of Spark.

The following is Spark Core's **Scala build tool** (**SBT**) dependency:

```
scalaVersion := "2.11.8"

libraryDependencies += "org.apache.spark" %% "spark-sql" % "2.4.0"
```

Spark Version 2.4.0 has introduced support for Scala Version 2.12; however, we will be using Scala Version 2.11 for exploring Spark's feature sets. Spark will be covered in more detail in the subsequent sections.

Spark setup and overview

Let's begin with the Spark setup on our local machine. Please refer to the Apache Spark official site (`https://spark.apache.org/downloads.html`) for details on downloading and installing Spark. At the time of this writing, Spark Version 2.4.0 is the latest version, so we will download and install this version. The following is a screenshot the Spark download web page:

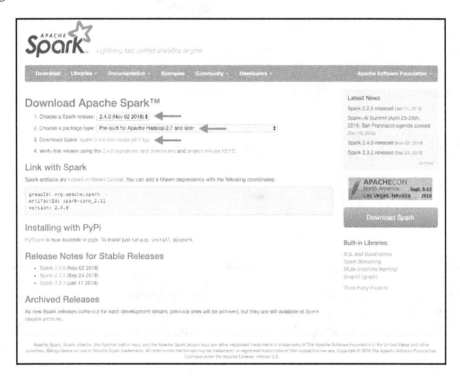

Please note the following pieces of information when selecting an appropriate download image:

- **Spark release**: Choose the latest stable release (2.4.0 at the time of this writing).
- **Package type**: Choose the **Pre-built for Apache Hadoop 2.7 and later** option.
- **Download link**: Clicking on this link will take you to the **Apache Download Mirrors** site. Select the site most suitable for you. Generally, the suggested mirror site works for most cases.

The following is another screenshot of the **Apache Download Mirrors** site with the Apache Spark image:

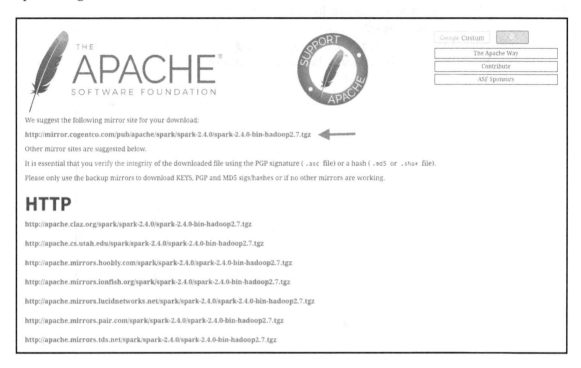

The downloadable image is a `.tar` file that is `.gzip` compressed. You can download the image by clicking the link in your browser. This should save the image in a `Download` directory. The name and location of the `Download` directory will generally vary from one OS to another. In the case of macOS, it is located in a subdirectory called `Downloads` located under the user's home directory.

Now, extract the Spark binaries by untaring the image file as follows. This is a macOS-specific example of extracting binaries:

```
$ tar -zxvf ~/Downloads/spark-2.4.0-bin-hadoop2.7.tgz
...
x spark-2.4.0-bin-hadoop2.7/README.md
x spark-2.4.0-bin-hadoop2.7/LICENSE
```

Alternatively, you could use the `curl` command to download the image as illustrated in the following code:

```
$ curl -o spark-2.4.0-bin-hadoop2.7.tgz
http://mirror.cogentco.com/pub/apache/spark/spark-2.4.0/spark-2.4.0-bin-had
```

```
oop2.7.tgz
  % Total % Received % Xferd Average Speed Time Time Time Current
  Dload Upload Total Spent Left Speed
100 217M 100 217M 0 0 14.1M 0 0:00:15 0:00:15 --:--:-- 14.2M
```

The preceding method uses the `curl` command to download the image and save it to the file named `spark-2.4.0-bin-hadoop2.7.tgz`. Next, you need to untar this file to extract the Spark binaries, as follows:

```
$ tar -zxvf spark-2.4.0-bin-hadoop2.7.tgz
...
x spark-2.4.0-bin-hadoop2.7/README.md
x spark-2.4.0-bin-hadoop2.7/LICENSE
```

At this point, there should be a directory called `spark-2.4.0-bin-hadoop2.7` created in your current directory. You can move this directory to any location of your choice.

Now, let's make sure that our Spark is set up correctly. Try the following set of commands in your shell to launch the Spark shell:

```
$ spark-2.4.0-bin-hadoop2.7/bin/spark-shell
2019-01-27 17:58:01 WARN NativeCodeLoader:62 - Unable to load native-hadoop
library for your platform... using builtin-java classes where applicable
Setting default log level to "WARN".
To adjust logging level use sc.setLogLevel(newLevel). For SparkR, use
setLogLevel(newLevel).
Spark context Web UI available at http://192.168.1.31:4040
Spark context available as 'sc' (master = local[*], app id =
local-1548640692711).
Spark session available as 'spark'.
Welcome to
      ____              __
     / __/__  ___ _____/ /__
    _\ \/ _ \/ _ `/ __/  '_/
   /___/ .__/\_,_/_/ /_/\_\   version 2.4.0
      /_/

Using Scala version 2.11.12 (Java HotSpot(TM) 64-Bit Server VM, Java
1.8.0_181)
Type in expressions to have them evaluated.
Type :help for more information.

scala>
```

It is important that the preceding command line produces output that is similar to this. The following are a few things to be noted:

- The Spark context Web UI is available at `http://192.168.1.31:4040` => where this is an URL (same as `http://localhost:4040`) a web UI where we can get more details about the current Spark session.
- The Spark context is available as `sc` (`master = local[*]`, `app id = local-1548640692711`) => We are running Spark in local mode.
- The Spark session is available as `'spark'` => the variable `'spark'` and provides us access to the Spark session.

The following is a screenshot of the UI (`http://localhost:4040`):

Spark UI is a powerful tool that helps us to understand how the Spark job works, and you can use it to get very useful insights into the different stages of execution. The landing page for this UI is `http://localhost:4040/jobs/`.

Please note that when you run the Spark shell, the following is created for you automatically:

- `Spark`: A `SparkSession` object that provides an entry point for interacting with Spark
- `sc`: A `SparkContext` object that provides an entry point for interacting with the Spark SQL

Let's see the features that are available in `SparkSession` and `sparkContext` by going through the following steps:

1. Start the Spark shell in your Terminal as follows:

```
$ spark-shell
```

2. Inside the Spark shell, check the type of Spark object, which must be an instance `SparkSession` for the `org.pache.spark.sql` package, as follows:

```scala
scala> spark.getClass
res0: Class[_ <: org.apache.spark.sql.SparkSession] = class
org.apache.spark.sql.SparkSession
```

3. Inside the Spark shell, type `spark. <TAB>` to get an insight into the methods and attributes of the `SparkSession` object, as follows:

```scala
scala> spark.
baseRelationToDataFrame conf emptyDataFrame implicits range
sessionState sql streams udf
catalog createDataFrame emptyDataset listenerManager read
sharedState sqlContext table version
close createDataset experimental newSession readStream sparkContext
stop time
```

The `SparkSession` provides a powerful set of APIs, which are very well documented here: `https://spark.apache.org/docs/2.4.0/api/java/org/apache/spark/sql/SparkSession.html`. To take full advantage of Spark, it is important to understand these APIs very well.

4. Import the Spark objects called `implicits`; these are automatically imported when a Spark shell is started. For a Spark session that is created by other mechanisms, these must be imported explicitly to take advantage of the implicit conversions, as follows:

```scala
scala> import spark.implicits._
import spark.implicits._
```

5. Make use of the Spark session's `implicits` to turn a `List` of integers to a Spark `Dataset`, as follows:

```scala
scala> val ds = List(1, 2, 3).toDS
ds: org.apache.spark.sql.Dataset[Int] = [value: int]
```

6. Check the `sc` type object. This must be an instance of `SparkContext` from the `org.apache.spark` package, as follows:

```scala
scala> sc.getClass
res1: Class[_ <: org.apache.spark.SparkContext] = class
org.apache.spark.SparkContext
```

7. Inside the Spark shell, type `sc. <TAB>` to get an insight into methods and attributes of `sparkContext` as follows:

```scala
scala> sc.
accumulable broadcast doubleAccumulator getSchedulingMode listJars
requestExecutors sparkUser
accumulableCollection cancelAllJobs emptyRDD hadoopConfiguration
longAccumulator requestTotalExecutors startTime
accumulator cancelJob files hadoopFile makeRDD runApproximateJob
statusTracker
addFile cancelJobGroup getAllPools hadoopRDD master runJob stop
addJar cancelStage getCheckpointDir isLocal newAPIHadoopFile
sequenceFile submitJob
addSparkListener clearCallSite getConf isStopped newAPIHadoopRDD
setCallSite textFile
appName clearJobGroup getExecutorMemoryStatus jars objectFile
setCheckpointDir uiWebUrl
applicationAttemptId collectionAccumulator getLocalProperty
killExecutor parallelize setJobDescription union
applicationId defaultMinPartitions getPersistentRDDs killExecutors
range setJobGroup version
binaryFiles defaultParallelism getPoolForName killTaskAttempt
register setLocalProperty wholeTextFiles
binaryRecords deployMode getRDDStorageInfo listFiles
removeSparkListener setLogLevel
```

The `SparkContext` also provides a powerful set of APIs, which are very well documented at: `https://spark.apache.org/docs/2.4.0/api/java/org/apache/spark/SparkContext.html`. The Spark context provides access to many of Spark's lower level features. For advanced users of Spark, it is important for you to understand these APIs very well.

8. Create a Spark RDD using `SparkContext`, as follows:

```scala
scala> val rdd = sc.parallelize(List(1, 2, 3))
rdd: org.apache.spark.rdd.RDD[Int] = ParallelCollectionRDD[0] at
parallelize at <console>:27
```

9. Compare and contrast the range API available in both `SparkSession` and `SparkContext` as follows:

```scala
scala> val rangeDS = spark.range(0, 10)
rangeDS: org.apache.spark.sql.Dataset[Long] = [id: bigint]

scala> val rangeRDD = sc.range(0, 10)
rangeRDD: org.apache.spark.rdd.RDD[Long] = MapPartitionsRDD[2] at
range at <console>:27
```

From `SparkSession`, we get a dataset of `Long`, whereas `SparkContext` returns an RDD of `Long` when using the range API. The dataset is a higher level Spark construct that is built on top of Spark's lower level RDD construct.

10. Stop the Spark session as follows:

```scala
scala> spark.stop()

scala>
```

11. Cleanly exit the Spark shell as follows:

```scala
scala> :quit
```

Spark core concepts

When our dataset is large, we can envision that there are multiple slices of data that make up the whole dataset. If a unit of compute work can be performed on each slice of data independently, then it is possible to parallelize this unit of computation, as shown in the following illustration:

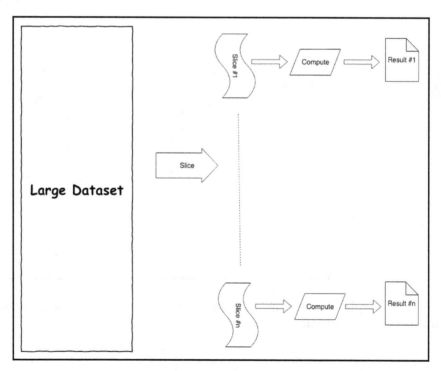

In the preceding illustration, there is a large dataset that gets sliced into multiple smaller slices. Identical computations run on each slice of data to create results from each slice.

At the core of Spark is the notion of RDD. Resilient implies that Spark is able to handle node failures automatically by retrying the compute work. Distributed, in this context, means that the dataset is spread across a cluster of nodes thereby overcoming the limitations associated with the resources of a single node.

Once we have this model of RDD in place, we can think of a slice of data and the computation together as one unit of work. This can be shipped to any worker to perform the work as illustrated in the following diagram:

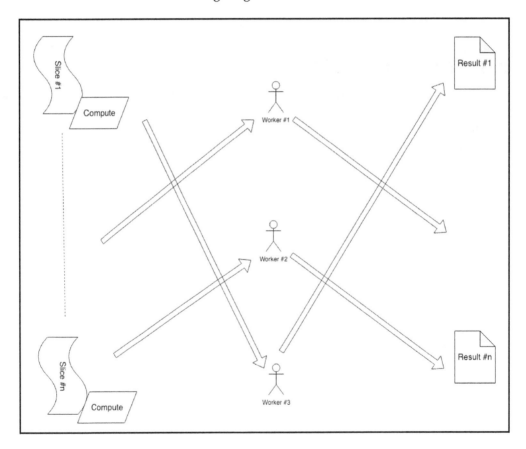

In the preceding example, we have three workers working on different slices of data and applying the associated compute.

Let's look at a concrete example to understand this concept further. Say we have a stack of cards with each card containing a number, and we want to find the maximum value from this stack of cards. In Scala REPL, we can solve this using the following steps:

1. Generate 20 random integers using Scala's random number generator, as follows:

```
// Generate 20 random ints
scala> val nums = for(i <- 1 to 20) yield scala.util.Random.nextInt
nums: scala.collection.immutable.IndexedSeq[Int] =
Vector(1701897084, -471839866, 289636030, -68368275, 1453521457,
1776989974, -333257299, 907038439, -157459682, 1279280488,
703554062, -658257712, 74262668, -2034769618, -1796054725,
1618075730, 642862982, 19687648, -1505425837, 1992429366)
```

2. Apply the `reduce` API on random numbers so that it picks up the higher of two numbers and applies this repeatedly to provide the highest number, as follows:

```
scala> nums.reduce((a, b) => if (a >= b) a else b) // reduce by
taking higher of the two values

res0: Int = 1992429366
```

3. Find the highest number using an alternative method, as follows:

```
// for illustration purpose, we will use the reduce method instead
scala> nums.max
res1: Int = 1992429366
```

In the preceding example, we used the `reduce` method from the Scala collection API to get the maximum value. The `reduce` method takes a function as an argument. The `supplied` function must accept two arguments whose types are the same as the element type of the collection. It must return a value that is also the same kind of element type. We can explore this in Scala REPL:

1. Define a Scala function that accepts two integers as input and returns an integer that is the higher of the two, as follows:

```
scala> val fun = (a: Int, b: Int) => if (a >= b) a else b
fun: (Int, Int) => Int = <function2>
```

2. Use the aforementioned function as a parameter to the reduce API of the collection to get the maximum value, as follows:

```
scala> nums.reduce(fun)
res2: Int = 1992429366
```

In the preceding example, we worked with only 20 numbers. If this is very large, a single-threaded operation would be quite slow to compute. The Scala collection API has support for parallelizing this compute. Let's explore this with one million numbers:

1. Create a million random integers as follows:

```
scala> val nums = for(i <- 1 to 1024*1024) yield
scala.util.Random.nextInt
nums: scala.collection.immutable.IndexedSeq[Int] =
Vector(357619961, 1737020067, -469045738, -601249939, -403302690,
-2066886866, -1785453571, -1547877670, -1485755408, 1037008188,
597778092, -11773505, -1087522271, -1065953174, -1910311733,
2031863519, -2077923104, 839563816, 1282957796, 674409356,
1813034923, -2070250813, -533697263, -1797217719, -751180312,
-1115480418, 890799862, -1566443600, -940178443, 1942197186,
1208980209, -1936454251, -1233813123, 1696121754, 882872208,
-1607840660, -1193358067, -249398026, 27578947, -1040824601,
62576870, 241072729, 914410066, -530844701, -1092314860,
1708591216, -2017362160, 1647649412, 1151979199, -197717793,
1392917841, -638219106, 2094838976, 567119171, 1904027672,
-216847530, -310681225, 1126606452, 1440522388, -1249070584,
1334505947, -...
```

2. Define the function that computes the higher of two integers as follows:

```
scala> val fun = (a: Int, b: Int) => if (a >= b) a else b
fun: (Int, Int) => Int = <function2>
```

3. Turn the collection into a parallel one and then apply the reduce API using the preceding function as the parameter, as follows:

```
scala> nums.par.reduce(fun) // turns into parallel collection and
then reduces
res4: Int = 2147483017
```

This worked for one million numbers. We can still go with a higher count of numbers; however, at some point, we will start seeing errors like those shown in the following code:

```
java.lang.OutOfMemoryError: GC overhead limit exceeded
  at java.lang.Integer.valueOf(Integer.java:832)
  at scala.runtime.BoxesRunTime.boxToInteger(BoxesRunTime.java:65)
  at $anonfun$1.apply(<console>:11)
  at
scala.collection.TraversableLike$$anonfun$map$1.apply(TraversableLike.scala
:234)
  at
scala.collection.TraversableLike$$anonfun$map$1.apply(TraversableLike.scala
:234)
```

```
at scala.collection.immutable.Range.foreach(Range.scala:160)
at scala.collection.TraversableLike$class.map(TraversableLike.scala:234)
at scala.collection.AbstractTraversable.map(Traversable.scala:104)
... 24 elided
```

In the preceding example, we are reaching the resource limits of a single machine. A machine has the following two key resources that affect the overall compute:

- **RAM**: Random access memory, where data is stored for computation
- **CPU core**: This performs the compute

For an approach that relies on a single machine for computations, the only option is to add more resources as the data volume grows. This approach reaches its limits fairly quickly, and in fact, the cost of such an approach gets fairly high with the addition of more resources.

Spark's RDD addresses this issue in a scalable way. At the core, RDD has the following two salient features:

- **Resilient**: RDDs preserve the dataset's consistency in the event of failures.
- **Distributed**: RDDs overcome the limitations of a single machine by distributing the dataset in a cluster of nodes.

Let's look at the same example in the Spark shell by going through the following steps:

1. Start the Spark shell as follows:

```
$ spark-shell
```

2. Create a Spark RDD of one million random integers as follows:

```
scala> val numRDD = spark.range(1024*1024).rdd.map(i =>
scala.util.Random.nextInt())
numRDD: org.apache.spark.rdd.RDD[Int] = MapPartitionsRDD[5] at map
at <console>:23
```

3. Get the number of partitions in RDD as follows:

```
scala> numRDD.getNumPartitions
res0: Int = 8
```

4. Print the size of each partition as follows:

```
scala> numRDD.foreachPartition(p => println(p.size))
131072
131072
131072
```

```
131072
131072
131072
131072
131072
```

5. Use the reduce API on the RDD compute maximum as follows:

```scala
scala> numRDD.reduce((a, b) => if (a >= b) a else b)
res2: Int = 2147483447
```

Let's look at what was being done here:

- `spark.range(1024*1024).rdd.map(i => scala.util.Random.nextInt())`:
    - We used Spark's `range` function to generate 1 million numbers and convert them to an RDD
    - We used RDD's `map` function to generate a random integer for each number
- `numRDD.getNumPartitions`:
    - There are eight partitions in this RDD
- `numRDD.foreachPartition(p => println(p.size))`:
    - Each partition of the RDD has 131,072 records
- `numRDD.reduce((a, b) => if (a >= b) a else b)`:
    - We used RDD's `reduce` function to get the maximum value
    - RDD's reduce API is similar to Scala collection's reduce API; however, RDD's reduce works on distributed data

Let's look at this in more detail by going to the Spark UI at `http://localhost:4040`:

We can see the following two jobs associated with the RDD:

- **Job #0**: The `foreach` operation on RDD that prints the size of each partition
- **Job #1**: The `reduce` operation on RDD that computes the maximum value

If we drill down further in **Job 1**, we can see the following details:

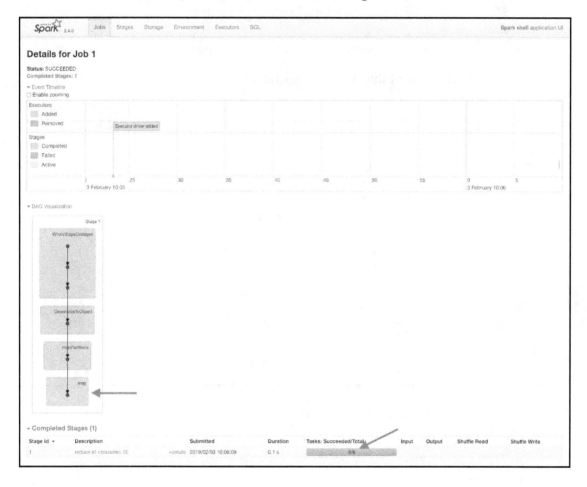

There was a `map` operation performed to randomize the numbers. As expected, there were in total eight tasks, since there were eight partitions in the RDD. The following are a few terms to note here:

- **Job**: A higher level unit of work that consists of one or more transformations and an action
- **Stage**: A series of transformations that happen within a job and are confined to a single partition
- **Task**: The work to be performed on a single partition of data

We can see more details about this in the **Stages** tab of the Spark UI, as shown in the following screenshot:

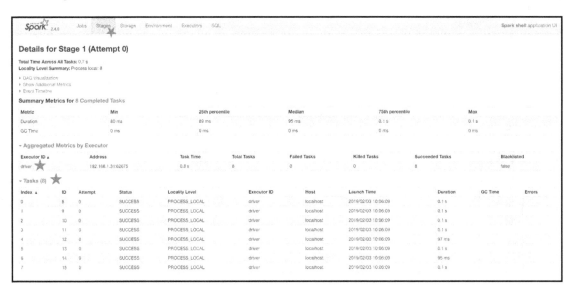

Spark uses lazy evaluation to perform its work. Spark's operations can be divided into the following two parts:

- **Transformations**: Operations that perform some data manipulations, data filtering
- **Actions**: Operations that materialize the results

When we perform transformations on an RDD in Spark, we are essentially building a recipe. When an action is performed, the recipe is materialized and action results are produced.

In the following code, we have created an RDD and performed a map transformation. At this point, RDD is defined; however, Spark has not performed any work, as seen in the following:

```
val numRDD = spark.range(1024*1024).rdd.map(i =>
scala.util.Random.nextInt())
```

Once we run the `reduce` operation, which is a Spark action, Spark starts to execute the following recipe that we defined earlier:

```
numRDD.reduce((a, b) => if (a >= b) a else b)
```

The reduce action creates a Spark job. This job consists of a single stage. Within this stage, there are eight tasks, one for each partition.

This lazy evaluation approach offers the following benefits:

- When a Spark action is executed, it looks at the entire execution graph that is needed to materialize the action. This provides opportunities for transformation optimization, such as eliminating redundant transformations and reordering operations that provide better overall performance while maintaining the overall consistency of the results produced.
- It provides opportunities for caching only the results that are used repeatedly. This becomes important when data to be handled is large.

Next, we will look at Spark's datasets and DataFrames while still exploring more details of Spark.

Spark Datasets and DataFrames

In the previous section, we looked at Spark's core functionality using RDDs. RDDs are powerful constructs; however, there are still some low-level details that a Spark user has to understand and master before making use of it. Spark's Datasets and DataFrame constructs provide higher level APIs for working with data.

Spark's Dataset brings a declarative style of programming along with the functional programming style of RDD. **Structured Query Language (SQL)** is a very popular declarative language, and is extremely popular among people who do not have a strong background in functional programming. The Spark DataFrame is a special type of dataset that provides the concepts of the row and column, as seen in the tradition **relational database (RDBS)** work.

Let's explore the example we used earlier using RDD. We will use the dataset and DataFrame constructs instead:

1. Start a `spark-shell` as follows:

   $ spark-shell

2. Create a dataset of one million random integers, as follows:

   ```
   scala> val numDS = spark.range(1024*1024).map(i =>
   scala.util.Random.nextInt())
   numDS: org.apache.spark.sql.Dataset[Int] = [value: int]
   ```

3. Use the reduce API of the dataset to compute the maximum, as follows:

   ```
   scala> numDS.reduce((a, b) => if (a >= b) a else b)
   res4: Int = 2147478392
   ```

4. Create a Spark DataFrame consisting of one million random integers, as follows:

   ```
   scala> val numDF = spark.range(1024*1024).map(i =>
   scala.util.Random.nextInt()).toDF
   numDF: org.apache.spark.sql.DataFrame = [value: int]
   ```

5. Try to perform reduce on the DataFrame; we should get an error, as follows:

   ```
   scala> numDF.reduce((a, b) => if (a >= b) a else b) // DOES NOT
   WORK
   <console>:26: error: value >= is not a member of
   org.apache.spark.sql.Row
            numDF.reduce((a, b) => if (a >= b) a else b)
   ```

We constructed a dataset and a DataFrame of random numbers instead of the RDD. We are able to perform a `reduce` action on the dataset to get the maximum value; however, the `reduce` action of the DataFrame produces an error. This is because each element of the DataFrame is of the `Row` type and so, the following operation involving two elements is incorrect:

```
(a, b) => if (a >= b) a else b
```

We can make the following modifications to produce the desired results:

```
scala> numDF.reduce((a, b) => if (a(0).asInstanceOf[Int] >=
b(0).asInstanceOf[Int]) a else b)
res8: org.apache.spark.sql.Row = [2147480464]
```

This DataFrame's `Row` has only one column, and it is of the `Int` type. We take the first column (at index 0) and cast it as `Int` before comparing the two values, as follows:

```
a(0).asInstanceOf[Int] // column at index 0 cast as an Int
```

Let's look at another concrete example to understand Spark's dataset and DataFrame properties in a Spark shell:

1. Start a Spark shell as follows:

   ```
   $ spark-shell
   ```

2. Define a Scala `case` class called `Person` as follows:

   ```
   scala> case class Person(fname: String, lname: String, age: Int)
   defined class Person
   ```

3. Create a small `List` of `Person` objects as follows:

   ```
   scala> val persons = List(Person("Jon", "Doe", 21), Person("Bob",
   "Smith", 25), Person("James", "Bond", 47))
   persons: List[Person] = List(Person(Jon,Doe,21),
   Person(Bob,Smith,25), Person(James,Bond,47))
   ```

4. Create a dataset of `Person` from the list of persons as follows:

   ```
   scala> val ds = spark.createDataset(persons)
   ds: org.apache.spark.sql.Dataset[Person] = [fname: string, lname:
   string ... 1 more field]
   ```

5. Create a DataFrame from the list of persons as follows:

   ```
   scala> val df = spark.createDataFrame(persons)
   df: org.apache.spark.sql.DataFrame = [fname: string, lname: string
   ... 1 more field]
   ```

6. Print a schema of the dataset as follows:

   ```
   scala> ds.printSchema
   root
    |-- fname: string (nullable = true)
    |-- lname: string (nullable = true)
    |-- age: integer (nullable = false)
   ```

7. Print a schema of the DataFrame as follows:

   ```
   scala> df.printSchema
   root
   ```

```
|-- fname: string (nullable = true)
|-- lname: string (nullable = true)
|-- age: integer (nullable = false)
```

In the preceding example, we did the following:

- Defined a case class called `Person` with three attributes:
  - `fname`: String
  - `lname`: String
  - `age`: Int
- Created a list of three persons
- Created a dataset of persons
- Created a DataFrame of persons
- Printed a schema of the dataset and the DataFrame

As discussed in `Chapter 1`, *Scala Overview*, Scala's case classes come in really handy when we work with Spark's datasets and, to some degree, DataFrames as well. Datasets provides a strong type safety that aids significantly in building robust data pipelines. The type of dataset in the previous example, it was `org.apache.spark.sql.Dataset[Person]`. This implies that it is a dataset of `Person`. When working with DataFrames of this type, safety is not available because DataFrame is a dataset of `Row`. We can confirm this in a Spark shell as follows:

```
scala> df.getClass
res11: Class[_ <: org.apache.spark.sql.DataFrame] = class
org.apache.spark.sql.Dataset

scala>
df.isInstanceOf[org.apache.spark.sql.Dataset[org.apache.spark.sql.Row]]
res12: Boolean = true
```

Both dataset and DataFrame are very powerful constructs in Spark, each with its own strengths. When both are used together, these become a powerful means of working with data. Spark's Datasets are only available in JVM programming languages. This means that datasets can be used only in Scala and Java. Spark's DataFrames, on the other hand, are supported in Scala, Java, Python, and R.

Let's look at some of the dataset APIs, continuing with the same example:

1. Show from rows from the dataset as follows:

```
scala> ds.show
+-----+-----+---+
|fname|lname|age|
```

```
+-----+-----+---+
|  Jon|  Doe|  21|
|  Bob|Smith|  25|
|James|  Bond|  47|
+-----+-----+---+
```

2. Convert first name and last name to uppercase by applying a map operation to each element of the dataset, as follows:

```scala
scala> val dsUpper = ds.map(p => p.copy(p.fname.toUpperCase,
p.lname.toUpperCase))
dsUpper: org.apache.spark.sql.Dataset[Person] = [fname: string,
lname: string ... 1 more field]
```

3. Show the row from the uppercase mapped dataset as follows:

```scala
scala> dsUpper.show
+-----+-----+---+
|fname|lname|age|
+-----+-----+---+
|  JON|  DOE|  21|
|  BOB|SMITH|  25|
|JAMES|  BOND|  47|
+-----+-----+---+
```

In the preceding example, we first displayed the contents of the dataset using the `show` command, which typically displays up to 20 entries. We then performed a `map` operation on each `Person` object by converting the first name and last name to uppercase. Finally, we displayed the contents of the transformed dataset.

Let's go through some similar steps with the DataFrame in the Spark shell, as follows:

1. Show some rows from DataFrame as follows:

```scala
scala> df.show
+-----+-----+---+
|fname|lname|age|
+-----+-----+---+
|  Jon|  Doe|  21|
|  Bob|Smith|  25|
|James|  Bond|  47|
+-----+-----+---+
```

2. Perform a `map` operation on DataFrame to convert the first and last name to uppercase. This returns an instance of the dataset where the attributes are named _1, _2, and _3, as follows:

```scala
scala> val dfUpper = df.map(r =>
(r(0).asInstanceOf[String].toUpperCase,
r(1).asInstanceOf[String].toUpperCase, r(2).asInstanceOf[Int]))
dfUpper: org.apache.spark.sql.Dataset[(String, String, Int)] = [_1:
string, _2: string ... 1 more field]
```

3. Fix the attribute name issue as follows:

```scala
scala> val dfUpperWithName = df.map(r =>
(r(0).asInstanceOf[String].toUpperCase,
r(1).asInstanceOf[String].toUpperCase,
r(2).asInstanceOf[Int])).toDF("fname", "lname", "age")
dfUpperWithName: org.apache.spark.sql.DataFrame = [fname: string,
lname: string ... 1 more field]
```

4. Show some rows from the mapped DataFrame as follows:

```scala
scala> dfUpperWithName.show
+-----+-----+---+
|fname|lname|age|
+-----+-----+---+
|  JON|  DOE| 21|
|  BOB|SMITH| 25|
|JAMES| BOND| 47|
+-----+-----+---+
```

There are some key differences in how the DataFrame works compared to the dataset. The `map` API on DataFrame converts it to a dataset of (`String`, `String`, `Int`). The other difference is that the object available to the `map` function is of the `Row` type as opposed to `Person`. Different parts of `Person` need to be extracted from `Row` and type cast to their appropriate types. There is also the following alternative way to achieve the same results without the need for type casting:

```scala
scala> val dfUpperWithName = df.map(r => ((r.getString(0), r.getString(1),
r.getInt(2)))).toDF("fname", "lname", "age")
dfUpperWithName: org.apache.spark.sql.DataFrame = [fname: string, lname:
string ... 1 more field]
```

Filtering data on the dataset and DataFrame can be performed in the following way:

1. Filter the dataset for entries where the age is greater than 25 as follows:

```
scala> val dsAbove25 = ds.where($"age" > 25)
dsAbove25: org.apache.spark.sql.Dataset[Person] = [fname: string,
lname: string ... 1 more field]
```

2. Filter DataFrame for entries where the age is greater than 25 as follows:

```
scala> val dfAbove25 = df.where($"age" > 25)
dfAbove25: org.apache.spark.sql.Dataset[org.apache.spark.sql.Row] =
[fname: string, lname: string ... 1 more field]
```

3. Show the filtered dataset's contents as follows:

```
scala> dsAbove25.show
+-----+-----+---+
|fname|lname|age|
+-----+-----+---+
|James| Bond| 47|
+-----+-----+---+
```

4. Show the filtered DataFrame's contents as follows:

```
scala> dfAbove24.show
+-----+-----+---+
|fname|lname|age|
+-----+-----+---+
|James| Bond| 47|
+-----+-----+---+
```

We used the `where` API by specifying a filter condition `$"age" > 25` in both cases. In this context, `$"age"` represents the column in the dataset or DataFrame. We can add more conditions to the `where` clause using the following steps:

1. Use multiple conditions to filter the dataset as follows:

```
scala> val ds25Bob = ds.where($"age" === 25 && $"fname" === "Bob")
ds25Bob: org.apache.spark.sql.Dataset[Person] = [fname: string,
lname: string ... 1 more field]
```

2. Use multiple conditions to filter the DataFrame as follows:

```
scala> val df25Bob = df.where($"age" === 25 && $"fname" === "Bob")
df25Bob: org.apache.spark.sql.Dataset[org.apache.spark.sql.Row] =
[fname: string, lname: string ... 1 more field]
```

3. Show the filtered dataset's contents as follows:

```
scala> ds25Bob.show
+-----+-----+---+
|fname|lname|age|
+-----+-----+---+
|  Bob|Smith| 25|
+-----+-----+---+
```

4. Show the filtered DataFrame's contents as follows:

```
scala> df25Bob.show
+-----+-----+---+
|fname|lname|age|
+-----+-----+---+
|  Bob|Smith| 25|
+-----+-----+---+
```

Please note the usage of the triple equals sign (===). This is needed to indicate that it is a column compare because the standard double equals (==) compares two references and returns a Boolean value.

We can also use a free-form expression to perform filtering in `where` clauses by using the following steps:

1. Apply multiple conditions on the dataset using the following expression:

```
scala> val dsWhereFF = ds.where("age = 25 and fname = 'Bob'")
dsWhereFF: org.apache.spark.sql.Dataset[Person] = [fname: string,
lname: string ... 1 more field]
```

2. Apply the same conditions to DataFrame using the following expression:

```
scala> val dfWhereFF = df.where("age = 25 and fname = 'Bob'")
dfWhereFF: org.apache.spark.sql.Dataset[org.apache.spark.sql.Row] =
[fname: string, lname: string ... 1 more field]
```

3. Show the filtered dataset's contents as follows:

```
scala> dsWhereFF.show
+-----+-----+---+
|fname|lname|age|
+-----+-----+---+
|  Bob|Smith| 25|
+-----+-----+---+
```

4. Show the filtered DataFrame's contents as follows:

```scala
scala> dfWhereFF.show
+-----+-----+---+
|fname|lname|age|
+-----+-----+---+
|  Bob|Smith| 25|
+-----+-----+---+
```

In this example, we are able to achieve the same results by using an SQL such as `where` condition.

Using the `select` API, we can select specific columns from the dataset and DataFrame, as follows:

```scala
scala> ds.select("fname", "lname")
res34: org.apache.spark.sql.DataFrame = [fname: string, lname: string]

scala> df.select("fname", "lname")
res35: org.apache.spark.sql.DataFrame = [fname: string, lname: string]
```

Applying `select` to a dataset returns a DataFrame, whereas applying `select` to DataFrame returns a DataFrame. It is an important observation that some APIs on the dataset returns a DataFrame. Similar to the `select` API, the `selectExpr` API also returns a DataFrame. The `selectExpr` API is a powerful API because it also allows transformations to be performed on the dataset and DataFrame columns.

Sourcing data using Spark

Spark provides a mechanism to work with a variety of data sources and formats. It also has excellent support for integrating with the **Hadoop Distributed File System (HDFS)**, as well as several other popular storage systems, such as Amazon S3. In this section, we will focus on the variety of data sources and formats supported by Spark.

Parquet file format

Apache Parquet (https://parquet.apache.org/) is an open source project and defines the specifications of a columnar data storage format. This storage format is extremely popular in the big data world for the following reasons:

- It supports nested data structures, which is good because most real-world data fits more naturally into a nested structure.

- Being columnar storage, it has analytical workloads where only a subset of columns is used for analysis.The Parquet columnar storage format leads to more efficient data scans.
- Parquet data is stored in row groups, which allows it to be splittable, and at the same time, data compression can be applied at row-group level without compromising splittable properties. This is important because this is how Spark is able to parallelize the processing by forming multiple partitions of a given dataset.
- It uses interesting encoding algorithms based on the properties of data that help in optimizing storage and data retrieval.
- Statistics about the data are stored along with the data as metadata. These are utilized further to reduce the scans when data is read.
- The schema is stored along with the data as part of the metadata. From the Parquet data files, it is very easy to decipher the schema of stored data.

The Spark API commonly used to read and write a Parquet file is as follows:

```
spark.read.parquet(sourceLocation)
```

```
dataframe.write.parquet(destinationLocation)
```

The Spark session's `read` method provides a `DataFrameReader` object that can be used to read various types of formats. More details about reading Parquet can be found at: `https://spark.apache.org/docs/2.4.0/api/java/org/apache/spark/sql/DataFrameReader.html#parquet-java.lang.String...-`.

On the other hand, a DataFrame's `write` method returns a `DataFrameWriter` object that can be used to write data in various types of formats. More details about writing data in Parquet can be found at: `https://spark.apache.org/docs/2.4.0/api/java/org/apache/spark/sql/DataFrameWriter.html#parquet-java.lang.String-`.

Both `DataFrameReader` and `DataFrameWriter` provide fairly comprehensive APIs to read and write data in many different formats.

Avro file format

Apache Avro (`https://avro.apache.org/`) is another data serialization format. This is a binary format that provides a compact representation of underlying data. Similar to Parquet, it is a structured data format and has support for storing nested data. Spark has excellent support for working with Avro.

Spark JDBC integration

A significant amount of enterprise data is stored in **relational database systems** (RDBMS). The majority of the more popular database systems support **Java Database Connectivity** (**JDBC**) as a way of interacting with these systems. Spark provides a convenient way to use JDBC for integrating with these RDBMS systems.

Using Spark to explore data

Spark's SQL provides a convenient way to explore data and gain a deeper understanding of the data. Spark's DataFrame construct can be registered as temporary tables. It is possible to run SQL on these registered tables by performing all of the normal operations, such as joining tables and filtering data.

Let's look at an example Spark shell to learn how to explore data by using the following steps:

1. Start the Spark shell in a Terminal as follows:

   ```
   $ spark-shell
   ```

2. Define the following Scala case called `Person` with the following three attributes:
   - `fname`: String
   - `lname`: String
   - `age`: Int

   ```
   scala> case class Person(fname: String, lname: String, age: Int)
   defined class Person
   ```

3. Create a Scala list consisting of a few persons and put it into a Spark dataset of `Person` as follows:

   ```
   scala> val personsDS = List(Person("Jon", "Doe", 22),
   Person("Jack", "Sparrow", 35), Person("James", "Bond", 47),
   Person("Mickey", "Mouse", 13)).toDS
   personsDS: org.apache.spark.sql.Dataset[Person] = [fname: string,
   lname: string ... 1 more field]
   ```

4. Create a Spark temporary view named `persons` with underlying coming from the dataset created in the previous step:

   ```
   scala> personsDS.createOrReplaceTempView("persons")
   ```

5. Run the SQL using the Spark session query in the temporary view created in the previous step. Limit the selection of persons to those aged 21 or older. This will return a new Spark DataFrame consisting of records that match the criteria. Please note that the object returned is a DataFrame, which is a special type of dataset of Row as follows:

```
scala> val personsAbove21 = spark.sql("select * from persons where
age >= 21")
personsAbove21: org.apache.spark.sql.DataFrame = [fname: string,
lname: string ... 1 more field]
```

6. Show the contents of the DataFrame created in the previous step as follows:

```
scala> personsAbove21.show(truncate=false)
+-----+-------+---+
|fname|lname  |age|
+-----+-------+---+
|Jon  |Doe    |22 |
|Jack |Sparrow|35 |
|James|Bond   |47 |
+-----+-------+---+
```

7. Run another SQL on the temporary view. Change the fname and lname fields to uppercase as follows:

```
scala> val personsUpperCase = spark.sql("select upper(fname) as
ufname, upper(lname) ulname, age from persons")
personsUpperCase: org.apache.spark.sql.DataFrame = [ufname: string,
ulname: string ... 1 more field]
```

8. Show the DataFrame created in the previous step as follows:

```
scala> personsUpperCase.show(truncate=false)
+------+-------+---+
|ufname|ulname |age|
+------+-------+---+
|JON   |DOE    |22 |
|JACK  |SPARROW|35 |
|JAMES |BOND   |47 |
|MICKEY|MOUSE  |13 |
+------+-------+---+
```

Summary

In this chapter, we explored the Apache Spark open source distributed data processing platform. We installed a copy of Apache Spark on our local computer. First, we learned about of Spark's core API using hands-on examples that explored Spark's **resilient distributed dataset (RDD)**. Next, we explored the higher level APIs of Spark using datasets and DataFrames.

In the next chapter, we will look at traditional machine learning concepts.

7
Traditional Machine Learning for Data Analysis

This chapter provides an overview of **machine learning** (**ML**) techniques for doing data analysis. In the previous chapters, we have explored some of the techniques that can be used by human beings to analyze and understand data. In this chapter, we look at how ML techniques could be used for similar purposes.

At the heart of ML is a number of algorithms that have proven to work for solving specific categories of problems with a high degree of effectiveness. This chapter covers the following popular ML methods:

- Decision trees
- Random forests
- Ridge and lasso regression
- k-means cluster analysis

It also covers the role of **natural language processing** (**NLP**) in effectively analyzing certain types of data problems. The discussion in this chapter is limited to traditional machine learning methods. It does not cover newer methods such as deep learning and neural networks.

Let's first get an understanding of ML before diving into various ML algorithms.

ML overview

Let's first look at what ML is. In a traditional sense, in order to solve a computational problem, we typically write explicit computer instructions that solve the problem based on all of the possible scenarios. The assumption here is that all of the rules associated with the specific problem being solved are known and well-defined in advance and could be codified into computer instructions. This assumption, however, is not always true. There are times when the rules are not known in advance and it is impractical to define deterministic rules that could be applied to solve the problem.

Let's look at this problem using a concrete example of an app stores where a consumer has the option of buying an app from a fairly large catalog of available apps. When the consumer logs into the app store, it displays a set of recommended apps that the consumer is highly likely to buy. The computational problem we have to solve here is to determine a small number of the most relevant apps that the consumer is highly likely to buy. Let's assume that we have the demographic information of the consumer, such as their age, sex, and location. We also have knowledge of the prior purchases of the consumer from the same app store. Writing explicit code to solve this problem for each individual consumer is going to be impractical and ineffective. Each time a new consumer is added to the app store, we would have to repeat this process. As the number of consumers grows, this methodology would turn out to be extremely expensive and not feasible to sustain. On the other hand, if we look at the buying behavior of a sufficiently large population of consumers in the app store over time and are able to establish some rules, we should be able to predict the likely purchases of the new consumer more effectively.

In the preceding example, we could use an ML algorithm that can sift through past observations of consumer purchases from the app store and then deduce some rules that could be used for predicting what apps a consumer is likely to purchase. These rules are sometimes known as an **analytical model**. The model could be fairly simple or complex based on the ML algorithm used and the richness of the data available.

The important property to observe here is that the machine learns from existing data by applying a specific algorithm to build an analytical model. Of course, the ML algorithm needs to be codified. However, the actual analytical model that gets built is based on the properties of data that the algorithm operates on. Here, the machine is learning about the properties of data using a specific algorithm, ultimately to arrive at a model. This model could be applied to make predictions about new and previously unseen input data points.

Characteristics of ML

The following are some of the most important characteristics of ML:

- The goal of ML is to build an analytical model without much human intervention.
- The machine analyzes already available-data by applying a specific ML algorithm and learning from the data. Human effort is still necessary for selecting an appropriate algorithm, determining the effectiveness of the algorithm, and tuning it up.
- The analytical model built by ML should be effective in making effective predictions about new and previously unseen data points. A machine is considered to be learning from data only if the built analytical model turns out to be effective in making future predictions.
- ML plays an important role in automating the process of model building.

Put simply, ML is the process of building an autonomous analytical model by applying an ML algorithm on already observed data.

Categories or types of ML

ML can be broken down into two distinct categories of learning:

- **Supervised learning**: The training data is labeled; that is, the training data consists of inputs as well as associated observed outcomes or outputs.
- **Unsupervised learning**: The training data is not labeled; that is, the training data consists of inputs only and there are no observed outcomes or outputs associated with it.

Next, we will look at some important ML algorithms from both of these categories.

Decision trees

As the name suggests, decision trees in ML build a tree-like structure with decision conditions on each branch. Conditions define the flow of the decision-making process. We can also think of decision trees as being similar to flow charts.

Decision trees are supervised ML algorithms. This implies that this algorithm learns from labeled data. It can be used for classification as well as regression.

Implementing decision trees

Let's look at a simple example to understand and explore this concept. We have the following observations:

Age in Years	Height in Inches	Weight in Pounds	Gender	Shoe Size
25	180	200	M	12
35	165	190	F	9
20	175	195	M	11
70	170	200	M	9
75	170	170	F	8

Our inputs, variables, or features are as follows:

- **Age in Years**
- **Height in Inches**
- **Weight in Pounds**
- **Gender**

The outcome or output is as follows:

- **Shoe Size**

The aforementioned is an example of labeled data. We can envision labeled data as a tuple of inputs and the corresponding output.

In this example, the problem that we are trying to solve is to predict the shoe size of a person based on their age, height, weight, and gender.

Using the decision trees ML algorithm, we can solve this problem, provided we have enough high-quality labeled training data. The algorithm will build a model based on the learnings from the training dataset. This model needs to be evaluated for effectiveness and this can be done by testing against labeled testing data.

From our intuitive understanding, we already know that for determining the shoe size of an adult, gender and height are the two most dominant factors. In the decision-making process, gender can also act well as a good split. The decision tree algorithm is about finding those good splits based on the provided labeled observations. This is a greedy algorithm that works recursively to find splits and then evaluates these splits to find the most effective splits.

Decision tree algorithms

Decision trees algorithms are of two types:

- **Classification**: Answering a yes/no type of question
- **Regression**: Outputting a continuous value

Implementing decision tree algorithms in our example

The example we are discussing here falls into the category of regression because our output is Shoe Size, which has multiple values.

If we ask a slightly different question, we could turn this into a classification problem. For example, given the following input variables, we can predict gender:

- `Age in Years`
- `Height in Inches`
- `Weight in Pounds`
- `Shoe Size`

In this simple example, we only have two possible values for this answer.

Let's look at how to run a decision tree classifier using Spark and Scala. The complete Spark example code is located here on GitHub:

```
https://github.com/apache/spark/blob/master/examples/src/main/scala/org/apache/
spark/examples/mllib/DecisionTreeClassificationExample.scala
```

Additionally, the following Spark documentation provides a good overview of the decision tree and the preceding example:

```
https://spark.apache.org/docs/2.4.0/mllib-decision-tree.html
```

The sample data for this example is located here:

```
https://github.com/apache/spark/blob/master/data/mllib/sample_libsvm_data.txt
```

For this exploration, it is best to download this file locally using the following `curl` command:

```
$ curl
    "https://raw.githubusercontent.com/apache/spark/master/data/mllib
        /sample_libsvm_data.txt" -o sample_libsvm_data.txt
```

It is important that while downloading, the raw link is used (`https://raw.` `githubusercontent.com/apache/spark/master/data/mllib/sample_libsvm_data.txt`) and the data is saved into a local file.

Run the following command in the shell to make sure that the downloaded data is correct:

```
$ head -1 sample_libsvm_data.txt
0 128:51 129:159 130:253 131:159 132:50 155:48 156:238 157:252 158:252
159:252 160:237 182:54 183:227 184:253 185:252 186:239 187:233 188:252
189:57 190:6 208:10 209:60 210:224 211:252 212:253 213:252 214:202 215:84
216:252 217:253 218:122 236:163 237:252 238:252 239:252 240:253 241:252
242:252 243:96 244:189 245:253 246:167 263:51 264:238 265:253 266:253
267:190 268:114 269:253 270:228 271:47 272:79 273:255 274:168 290:48
291:238 292:252 293:252 294:179 295:12 296:75 297:121 298:21 301:253
302:243 303:50 317:38 318:165 319:253 320:233 321:208 322:84 329:253
330:252 331:165 344:7 345:178 346:252 347:240 348:71 349:19 350:28 357:253
358:252 359:195 372:57 373:252 374:252 375:63 385:253 386:252 387:195
400:198 401:253 402:190 413:255 414:253 415:196 427:76 428:246 429:252
430:112 441:253 442:252 443:148 455:85 456:252 457:230 458:25 467:7 468:135
469:253 470:186 471:12 483:85 484:252 485:223 494:7 495:131 496:252 497:225
498:71 511:85 512:252 513:145 521:48 522:165 523:252 524:173 539:86 540:253
541:225 548:114 549:238 550:253 551:162 567:85 568:252 569:249 570:146
571:48 572:29 573:85 574:178 575:225 576:253 577:223 578:167 579:56 595:85
596:252 597:252 598:252 599:229 600:215 601:252 602:252 603:252 604:196
605:130 623:28 624:199 625:252 626:252 627:253 628:252 629:252 630:233
631:145 652:25 653:128 654:252 655:253 656:252 657:141 658:37
```

This data is created as per **Library for Support Vector Machines** (**LIBSVM**) specifications and more details about LIBSVM can be found at `https://www.csie.ntu.edu.tw/~cjlin/` `libsvmtools/datasets/`.

Now, let's build a decision tree classifier using this dataset in `spark-shell`:

1. Start `spark-shell` in your Terminal:

   ```
   $ spark-shell
   ```

2. Import `DecisionTree` from Spark's MLLib's `tree` package. More information about this class can be found at `https://spark.apache.org/docs/2.4.1/api/` `java/index.html?org/apache/spark/mllib/tree/DecisionTree.html`. This class implements a decision tree learning algorithm for classification and regression:

   ```scala
   scala> import org.apache.spark.mllib.tree.DecisionTree
   import org.apache.spark.mllib.tree.DecisionTree
   ```

3. Import `DecisionTreeModel` from Spark's MLLib's `tree.model` package. More information about this can be found at https://spark.apache.org/docs/2.4.1/api/java/index.html?org/apache/spark/mllib/tree/DecisionTree.html:

```scala
scala> import org.apache.spark.mllib.tree.model.DecisionTreeModel
import org.apache.spark.mllib.tree.model.DecisionTreeModel
```

4. Import `MLUtils` from Spark's MLLib's `util` package. More information about this can be found at https://spark.apache.org/docs/2.4.1/api/java/index.html?org/apache/spark/mllib/tree/DecisionTree.html. `MLUtils` provides utility methods for loading, saving, and preprocessing data used in MLLib:

```scala
scala> import org.apache.spark.mllib.util.MLUtils
import org.apache.spark.mllib.util.MLUtils
```

5. Using `MLUtils`, load the sample data in LIBSVM format. Note that doing so returns a Spark RDD of MLLib's `LabeledPoint`:

```scala
scala> val data = MLUtils.loadLibSVMFile(sc,
"sample_libsvm_data.txt") // load data in LIBSVM format
data:
org.apache.spark.rdd.RDD[org.apache.spark.mllib.regression.LabeledP
oint] = MapPartitionsRDD[6] at map at MLUtils.scala:86
```

6. Split the data randomly into two parts: one with 70% of the data and another one with the remaining 30%. Note that we are using Spark RDD's `randomSplit` API to do so. This provides us with an RDD array of `LabeledPoint`:

```scala
// 70/30 split
scala> val splits = data.randomSplit(Array(0.7, 0.3))
splits:
Array[org.apache.spark.rdd.RDD[org.apache.spark.mllib.regression.La
beledPoint]] = Array(MapPartitionsRDD[7] at randomSplit at
<console>:28, MapPartitionsRDD[8] at randomSplit at <console>:28)
```

7. Designate the 70% split as training data and the 30% split as test data. This is a fairly standard practice to reserve 30% of the labeled data for testing. We now have two RDDs of `LabeledPoint`, training and test:

```scala
// 70% is training data and 30% test data
scala> val (trainingData, testData) = (splits(0), splits(1))
trainingData: org.apache.spark.rdd.RDD
                [org.apache.spark.mllib.regression.LabeledPoint] =
                    MapPartitionsRDD[7] at randomSplit at <console>:28
testData:
org.apache.spark.rdd.RDD[org.apache.spark.mllib.regression
```

```
.LabeledPoint] = MapPartitionsRDD[8] at randomSplit at
<console>:28
```

8. Define the number of classes. In this case, the value is 2 because only two values are possible:

```
scala> val numClasses = 2 // two possible values (0 or 1)
numClasses: Int = 2
```

9. Define an empty `Map` method for categorical features because there are no categorical features for this use case. Note that the key and value for the `Map` are both of type `Int`:

```
scala> val categoricalFeaturesInfo = Map[Int, Int]() // no
categorical features
categoricalFeaturesInfo: scala.collection.immutable.Map[Int,Int] =
Map()
```

10. Define the impurity as a `gini` impurity measure:

```
scala> val impurity = "gini"
impurity: String = gini
```

11. Define the maximum depth of the tree:

```
scala> val maxDepth = 5 // maximum depth of the tree
maxDepth: Int = 5
```

12. Define the maximum width of the tree:

```
scala> val maxBins = 32 // maximum with of the tree
maxBins: Int = 32
```

13. Create a decision tree model by training on the training data and using the parameter previously specified:

```
scala> val model = DecisionTree.trainClassifier(trainingData,
numClasses, categoricalFeaturesInfo, impurity, maxDepth, maxBins)
model: org.apache.spark.mllib.tree.model.DecisionTreeModel =
DecisionTreeModel classifier of depth 2 with 5 nodes
```

14. Output the model as follows:

```
scala> model.toDebugString
res0: String =
"DecisionTreeModel classifier of depth 2 with 5 nodes
  If (feature 351 <= 38.0)
    If (feature 125 <= 254.5)
```

```
        Predict: 0.0
      Else (feature 125 > 254.5)
        Predict: 1.0
    Else (feature 351 > 38.0)
      Predict: 1.0
  "
```

Let's take a look at the data:

```
0 128:51 129:159 130:253 131:159 132:50 155:48 156:238 157:252 158:252 ....
```

Here is what we can see:

- The first value is the outcome or result. This has two possible values: 0 or 1.
- Subsequent values are listed as `<feature id>:<associated value>`.

Evaluating the results

The decision tree training classifier is able to find a good split at features 351 and 125. Please note that you might see a completely different model because the training and test data is getting randomly split into a 70:30 ratio.

Let's evaluate how the model performs on the test data:

1. For each record in `testData`, first, predict the results using the features of the record by applying the model. Record this as an observed value and a predicted value pair:

    ```scala
    scala> val labelAndPreds = testData.map { point =>
         | val prediction = model.predict(point.features)
         | (point.label, prediction)
         | }
    labelAndPreds: org.apache.spark.rdd.RDD[(Double, Double)] =
    MapPartitionsRDD[24] at map at <console>:30
    ```

2. Calculate how many records were incorrectly predicted:

    ```scala
    scala> val testErr = labelAndPreds.filter(r => r._1 !=
    r._2).count().toDouble / testData.count()
    testErr: Double = 0.11764705882352941
    ```

We have a roughly 12% error in predicting the correct outcome.

3. This model can be saved to a file using the following API:

    ```scala
    scala> model.save(sc, "myDecisionTreeClassificationModel")
    ```

This saves the model to the `myDecisionTreeClassificationModel` directory.

4. Exit out of `spark-shell` as follows:

```
scala> spark.stop()

scala> :quit
```

Using our model with a decision tree

Now the model is ready to use:

1. Start the `spark-shell`:

```
$ spark-shell
```

2. Import `DecisionTreeModel` from Spark's MLLib package:

```
scala> import org.apache.spark.mllib.tree.model.DecisionTreeModel
import org.apache.spark.mllib.tree.model.DecisionTreeModel
```

3. Use `DecisionTreeModel` objects' load API to load the model:

```
scala> val model = DecisionTreeModel.load(sc,
"myDecisionTreeClassificationModel")
model: org.apache.spark.mllib.tree.model.DecisionTreeModel =
DecisionTreeModel classifier of depth 2 with 5 nodes
```

4. The output of the model is as follows:

```
scala> model.toDebugString
res0: String =
"DecisionTreeModel classifier of depth 2 with 5 nodes
  If (feature 351 <= 38.0)
   If (feature 125 <= 254.5)
    Predict: 0.0
   Else (feature 125 > 254.5)
    Predict: 1.0
  Else (feature 351 > 38.0)
    Predict: 1.0
"
```

We can see that the loaded model is the same one that we built earlier. This model can be used to predict outcomes of unseen observations by doing the following:

```
scala> model.predict(features) // features is a new observation
```

The model will output a value of 0 or 1.

Random forest

Random forest is an easy-to-use and powerful ML algorithm. It is also a supervised algorithm and requires labeled data to learn from. In fact, the decision tree acts as the building block for the random forest algorithm. Just like the decision tree, the random forest ML algorithm can be used for classification as well as regression.

The fundamental motivation behind the random forest algorithm is to combine results from multiple random decision trees into a single model. One very nice outcome of the random forest algorithm is that it prevents overfitting of the model to the training dataset.

Random forest algorithms

The random forest algorithm can be summarized as follows:

- Each decision tree in a random forest uses a subset of random features.
- Only a random subset of training data is used to build a decision tree.
- Models from multiple random decision trees are combined to build a single model.

We can think of a random forest algorithm as something like consensus building, where we take multiple, diverse ideas and converge these into a single idea.

In the real world, collective wisdom has proven to provide very effective guidance in multiple, diverse fields. The random forest algorithm essentially borrows from the same idea.

Let's explore the random forest algorithm using Spark ML. We will be using the same `sample_libsvm_data.txt` data file from the decision tree example:

1. Run Spark shell from your Terminal:

   ```
   $ spark-shell
   ```

2. Import the required classes from Spark's ML package:

   ```scala
   scala> import org.apache.spark.ml.Pipeline
   import org.apache.spark.ml.Pipeline
   scala> import
   org.apache.spark.ml.classification.{RandomForestClassificationModel
   ```

```
, RandomForestClassifier}
import
org.apache.spark.ml.classification.{RandomForestClassificationModel
, RandomForestClassifier}
scala> import
org.apache.spark.ml.evaluation.MulticlassClassificationEvaluator
import
org.apache.spark.ml.evaluation.MulticlassClassificationEvaluator
scala> import org.apache.spark.ml.feature.{IndexToString,
StringIndexer, VectorIndexer}
import org.apache.spark.ml.feature.{IndexToString, StringIndexer,
VectorIndexer}
```

3. Use SparkSession to read the `_libsvm_data.txt` file sample in `libsvm` format as a Spark DataFrame:

```
scala> val data =
spark.read.format("libsvm").load("./sample_libsvm_data.txt")
2019-04-24 21:49:27 WARN LibSVMFileFormat:66 - 'numFeatures' option
not specified, determining the number of features by going though
the input. If you know the number in advance, please specify it via
'numFeatures' option to avoid the extra scan.
data: org.apache.spark.sql.DataFrame = [label: double, features:
vector]
```

4. Index the label column of the source DataFrame and create a new DataFrame with `indexedLabel` as an additional column:

```
scala> val labelIndexer = new
StringIndexer().setInputCol("label").setOutputCol("indexedLabel").f
it(data)
labelIndexer: org.apache.spark.ml.feature.StringIndexerModel =
strIdx_484cb0e8e765
```

5. Index the features column of the source DataFrame and create a new DataFrame with `indexedFeatures` as an additional column. Make sure that the features are treated as continuous variables if there are more than four distinct values, otherwise, they will be categorical:

```
scala> val featureIndexer = new
VectorIndexer().setInputCol("features").setOutputCol("indexedFeatur
es").setMaxCategories(4).fit(data)
featureIndexer: org.apache.spark.ml.feature.VectorIndexerModel =
vecIdx_07847af33557
```

6. Randomly split the source DataFrame into two DataFrames using a 7:3 ratio. The first DataFrame is to be used for training purposes. The second DataFrame is to be used for testing purposes:

```scala
scala> val Array(trainingData, testData) =
data.randomSplit(Array(0.7, 0.3))
trainingData:
org.apache.spark.sql.Dataset[org.apache.spark.sql.Row] = [label:
double, features: vector]
testData: org.apache.spark.sql.Dataset[org.apache.spark.sql.Row] =
[label: double, features: vector]
```

7. Create `RandomForestClassifier` using the builder pattern. Set the label column name as `indexedLabel` and the features column name as `features column`. Allow 10 decision trees to be used for arriving at a consensus:

```scala
scala> val randomForest = new
RandomForestClassifier().setLabelCol("indexedLabel").setFeaturesCol
("indexedFeatures").setNumTrees(10)
randomForest:
org.apache.spark.ml.classification.RandomForestClassifier =
rfc_954d41674853
```

8. Create a label converter object that converts indexed labels back to the original labels:

```scala
scala> val labelConverter = new
IndexToString().setInputCol("prediction").setOutputCol("predictedLa
bel").setLabels(labelIndexer.labels)
labelConverter: org.apache.spark.ml.feature.IndexToString =
idxToStr_74f8a7c145d8
```

9. Create a `Pipeline` object that chains the label indexer, feature indexer, random forest, and label converter:

```scala
scala> val pipeline = new Pipeline().setStages(Array(labelIndexer,
featureIndexer, randomForest, labelConverter))
pipeline: org.apache.spark.ml.Pipeline = pipeline_8cd293d72ebd
```

10. Run the pipeline to fit training data and create a model. Doing so runs the label indexer, feature indexer, random forest, and label converter one after the other:

```scala
scala> val model = pipeline.fit(trainingData)
model: org.apache.spark.ml.PipelineModel = pipeline_8cd293d72ebd
```

11. Using the model, run `predictions` on the test data:

```
scala> val predictions = model.transform(testData)
predictions: org.apache.spark.sql.DataFrame = [label: double,
features: vector ... 6 more fields]
```

12. Display five rows from the created `predictions` DataFrame:

```
scala> predictions.select("predictedLabel", "label",
"features").show(5, true)
+--------------+-----+--------------------+
|predictedLabel|label|  features|
+--------------+-----+--------------------+
|  0.0|  0.0|(692,[95,96,97,12...|
|  0.0|  0.0|(692,[100,101,102...|
|  0.0|  0.0|(692,[122,123,148...|
|  0.0|  0.0|(692,[123,124,125...|
|  0.0|  0.0|(692,[124,125,126...|
+--------------+-----+--------------------+
only showing top 5 rows
```

13. Evaluate the accuracy of the predicted label by comparing it against the original label:

```
scala> val evaluator = new
MulticlassClassificationEvaluator().setLabelCol("indexedLabel").set
PredictionCol("prediction").setMetricName("accuracy")
evaluator:
org.apache.spark.ml.evaluation.MulticlassClassificationEvaluator =
mcEval_4b737dba9096

scala> val accuracy = evaluator.evaluate(predictions)
accuracy: Double = 1.0

scala> println(s"Test Error = ${(1.0 - accuracy)}")
Test Error = 0.0
```

14. Extract the built model, getting the element at index 2 of the pipeline stages:

```
scala> val rfModel =
model.stages(2).asInstanceOf[RandomForestClassificationModel]
rfModel:
org.apache.spark.ml.classification.RandomForestClassificationModel
= RandomForestClassificationModel (uid=rfc_954d41674853) with 10
trees
```

15. Print out the contents of the built model:

```
scala> println(s"Learned classification forest model:\n
${rfModel.toDebugString}")
Learned classification forest model:
 RandomForestClassificationModel (uid=rfc_954d41674853) with 10
trees
  Tree 0 (weight 1.0):
    If (feature 552 <= 5.5)
     If (feature 441 <= 2.5)
      If (feature 207 <= 161.0)
       Predict: 0.0
      Else (feature 207 > 161.0)
       Predict: 1.0
     Else (feature 441 > 2.5)
      Predict: 1.0
    Else (feature 552 > 5.5)
     Predict: 1.0
  Tree 1 (weight 1.0):
    If (feature 463 <= 19.5)
     If (feature 317 <= 8.0)
      If (feature 544 <= 135.0)
       Predict: 1.0
      Else (feature 544 > 135.0)
       Predict: 0.0
     Else (feature 317 > 8.0)
      Predict: 1.0
    Else (feature 463 > 19.5)
     Predict: 0.0
  Tree 2 (weight 1.0):
    If (feature 540 <= 102.5)
     If (feature 101 <= 57.0)
      Predict: 0.0
     Else (feature 101 > 57.0)
      Predict: 1.0
    Else (feature 540 > 102.5)
     Predict: 1.0
  Tree 3 (weight 1.0):
    If (feature 328 <= 24.0)
     If (feature 690 <= 3.5)
      If (feature 481 <= 14.5)
       If (feature 100 <= 44.0)
        Predict: 0.0
       Else (feature 100 > 44.0)
        Predict: 1.0
      Else (feature 481 > 14.5)
       Predict: 1.0
     Else (feature 690 > 3.5)
```

```
        Predict: 1.0
     Else (feature 328 > 24.0)
     Predict: 1.0
  Tree 4 (weight 1.0):
    If (feature 429 <= 7.0)
     If (feature 407 <= 9.5)
      Predict: 1.0
     Else (feature 407 > 9.5)
      Predict: 0.0
    Else (feature 429 > 7.0)
     Predict: 1.0
  Tree 5 (weight 1.0):
    If (feature 462 <= 62.5)
     Predict: 1.0
    Else (feature 462 > 62.5)
     Predict: 0.0
  Tree 6 (weight 1.0):
    If (feature 512 <= 8.0)
     If (feature 289 <= 249.0)
      If (feature 550 <= 46.0)
       Predict: 0.0
      Else (feature 550 > 46.0)
       Predict: 1.0
     Else (feature 289 > 249.0)
      Predict: 1.0
    Else (feature 512 > 8.0)
     Predict: 1.0
  Tree 7 (weight 1.0):
    If (feature 512 <= 8.0)
     If (feature 288 <= 154.5)
      Predict: 0.0
     Else (feature 288 > 154.5)
      Predict: 1.0
    Else (feature 512 > 8.0)
     Predict: 1.0
  Tree 8 (weight 1.0):
    If (feature 462 <= 62.5)
     Predict: 1.0
    Else (feature 462 > 62.5)
     Predict: 0.0
  Tree 9 (weight 1.0):
    If (feature 377 <= 103.5)
     If (feature 435 <= 32.5)
      Predict: 1.0
     Else (feature 435 > 32.5)
      Predict: 0.0
    Else (feature 377 > 103.5)
     If (feature 317 <= 8.0)
```

```
Predict: 0.0
Else (feature 317 > 8.0)
  Predict: 1.0
```

Ridge and lasso regression

Ridge and lasso regression are supervised linear regression ML algorithms. Both of these algorithms aim at reducing model complexity and prevent overfitting. When there is a large number of features or variables in a training dataset, the model built by ML generally tends to be complex.

Characteristics of ridge regression

The key characteristics of ridge regression are as follows:

- **Coefficient shrinkage**: This helps in reducing model complexity
- **Regularization**: This adds information to prevent overfitting

Characteristics of lasso regression

Lasso stands for **least absolute shrinkage and selection operator**. The following are key characteristics of lasso regression:

- **Feature selection**: Selecting a subset of the most relevant features from a large number of features
- **Regularization**: Adding information to prevent overfitting

Let's look at running linear regression methods on Spark:

1. Download the data file needed for exploration in your Terminal:

   ```
   $ curl
   https://raw.githubusercontent.com/apache/spark/master/data/mllib/ri
   dge-data/lpsa.data -o lpsa.data
   ```

2. Start a Spark shell in your Terminal:

   ```
   $ spark-shell
   ```

3. Import the necessary classes from the Spark MLLib package, which are needed for linear regression:

```scala
scala> import org.apache.spark.mllib.linalg.Vectors
import org.apache.spark.mllib.linalg.Vectors
scala> import org.apache.spark.mllib.regression.LabeledPoint
import org.apache.spark.mllib.regression.LabeledPoint
scala> import
org.apache.spark.mllib.regression.LinearRegressionModel
import org.apache.spark.mllib.regression.LinearRegressionModel
scala> import
org.apache.spark.mllib.regression.LinearRegressionWithSGD
import org.apache.spark.mllib.regression.LinearRegressionWithSGD
```

4. Load the data as follows:

```scala
scala> val data = sc.textFile("./lpsa.data")
data: org.apache.spark.rdd.RDD[String] = ./lpsa.data
MapPartitionsRDD[1] at textFile at <console>:28
```

5. Parse the data into `LabeledPoint` and `cache`:

```scala
scala> val parsedData = data.map { line =>
     | val parts = line.split(',')
     | LabeledPoint(parts(0).toDouble,
Vectors.dense(parts(1).split(' ').map(_.toDouble)))
     | }.cache()
parsedData:
org.apache.spark.rdd.RDD[org.apache.spark.mllib.regression.LabeledP
oint] = MapPartitionsRDD[2] at map at <console>:29
```

6. Build the model by setting the number of step-size iterations:

```scala
scala> val numIterations = 100
numIterations: Int = 100

scala> val stepSize = 0.00000001
stepSize: Double = 1.0E-8
scala> val model = LinearRegressionWithSGD.train(parsedData,
numIterations, stepSize)
model: org.apache.spark.mllib.regression.LinearRegressionModel =
org.apache.spark.mllib.regression.LinearRegressionModel: intercept
= 0.0, numFeatures = 8
```

7. Evaluate the model on the training examples and compute the training errors, as follows:

```scala
scala> val valuesAndPreds = parsedData.map { point =>
```

```
    | val prediction = model.predict(point.features)
    | (point.label, prediction)
    | }
valuesAndPreds: org.apache.spark.rdd.RDD[(Double, Double)] =
MapPartitionsRDD[9] at map at <console>:31

scala> val MSE = valuesAndPreds.map{ case(v, p) => math.pow((v -
p), 2) }.mean()
MSE: Double = 7.4510328101026015
scala> println(s"training Mean Squared Error $MSE")
training Mean Squared Error 7.4510328101026015
```

8. Save the model as follows:

```
scala> model.save(sc, "./LinearRegressionWithSGDModel")
```

9. Load the saved model as follows:

```
scala> val sameModel = LinearRegressionModel.load(sc,
"./LinearRegressionWithSGDModel")
sameModel: org.apache.spark.mllib.regression.LinearRegressionModel
= org.apache.spark.mllib.regression.LinearRegressionModel:
intercept = 0.0, numFeatures = 8
```

10. Output the model as follows:

```
scala> sameModel
res2: org.apache.spark.mllib.regression.LinearRegressionModel =
org.apache.spark.mllib.regression.LinearRegressionModel: intercept
= 0.0, numFeatures = 8
scala> sameModel.toPMML
res3: String =
"<?xml version="1.0" encoding="UTF-8" standalone="yes"?>
<PMML version="4.2" xmlns="http://www.dmg.org/PMML-4_2">
    <Header description="linear regression">
        <Application name="Apache Spark MLlib" version="2.4.0"/>
        <Timestamp>2019-04-24T22:45:35</Timestamp>
    </Header>
    <DataDictionary numberOfFields="9">
        <DataField name="field_0" optype="continuous"
dataType="double"/>
        <DataField name="field_1" optype="continuous"
dataType="double"/>
        <DataField name="field_2" optype="continuous"
dataType="double"/>
        <DataField name="field_3" optype="continuous"
dataType="double"/>
        <DataField name="field_4" optype="continuous"
```

```
dataType="double"/>
        <DataField name="field_5" optype="continuous"
dataType="double"/>
        <...
```

In the preceding example, we used Spark MLLib's `LinearRegressionWithSGD` class for building a linear regression model. Spark MLLib has two other classes; `RidgeRegressionWithSGD` and `LassoWithSGD` can be used in a similar fashion to build linear models in Spark.

k-means cluster analysis

k-means is a clustering ML algorithm. This is a nonsupervised ML algorithm. Its primary use is for clustering together closely related data and gaining an understanding of the structural properties of the data.

As the name suggests, this algorithm tries to form a k number of clusters around k-mean values. How many clusters are to be formed, that is, the value of k, is something a human being has to determine at the outset. This algorithm relies on the Euclidean distance to calculate the distance between two points. We can think of each observation as a point in n-dimensional space, where n is the number of features. The distance between two observations is the Euclidean distance between these in n-dimensional space.

To begin with, the algorithm picks up k random records from the dataset. These are the initial k-mean values. In the next step, for each record in the set, it calculates the Euclidean distance of this record from each k-mean record and assigns the record to the k-mean record with the lowest Euclidean distance. At the end of this, it has the first k clusters. Now for each cluster, it computes the mean value and thus has k-mean values. These k-mean values are used for the next iteration of the algorithm. It repeats these steps with these k values and arrives at a new step of k values, which are used for the next iteration of the algorithm. This gets repeated over and over again until the k-mean values start to converge, that is, k centroids are established around where the data is concentrated.

Let's explore k-means clustering using Spark's MLLib library:

1. Download the test data in your Terminal using the following command:

```
$ curl
https://raw.githubusercontent.com/apache/spark/master/data/mllib/km
eans_data.txt -o kmeans_data.txt
```

2. Start Spark shell in your Terminal:

```
$ spark-shell
```

3. Import the necessary classes from Spark's MLLib package:

```
scala> import org.apache.spark.mllib.clustering.{KMeans,
KMeansModel}
import org.apache.spark.mllib.clustering.{KMeans, KMeansModel}
scala> import org.apache.spark.mllib.linalg.Vectors
import org.apache.spark.mllib.linalg.Vectors
```

4. Load and parse the data:

```
scala> val data = sc.textFile("./kmeans_data.txt")
data: org.apache.spark.rdd.RDD[String] = ./kmeans_data.txt
MapPartitionsRDD[1] at textFile at <console>:26
scala> val parsedData = data.map(s => Vectors.dense(s.split('
').map(_.toDouble))).cache()
parsedData:
org.apache.spark.rdd.RDD[org.apache.spark.mllib.linalg.Vector] =
MapPartitionsRDD[2] at map at <console>:27
```

5. Cluster the data into two classes using `KMeans`:

```
scala> val numClusters = 2
numClusters: Int = 2
scala> val numIterations = 20
numIterations: Int = 20
scala> val clusters = KMeans.train(parsedData, numClusters,
numIterations)
2019-04-24 22:24:17 WARN BLAS:61 - Failed to load implementation
from: com.github.fommil.netlib.NativeSystemBLAS
2019-04-24 22:24:17 WARN BLAS:61 - Failed to load implementation
from: com.github.fommil.netlib.NativeRefBLAS
clusters: org.apache.spark.mllib.clustering.KMeansModel =
org.apache.spark.mllib.clustering.KMeansModel@77f43f3e
```

6. Evaluate the clustering by computing `Within Set Sum of Squared Errors`:

```
scala> val WSSSE = clusters.computeCost(parsedData)
WSSSE: Double = 0.11999999999994547
scala> println(s"Within Set Sum of Squared Errors = $WSSSE")
Within Set Sum of Squared Errors = 0.11999999999994547
```

7. Save the model:

```
scala> clusters.save(sc, "./KMeansModel")
```

8. Load the saved model:

```scala
scala> val sameModel = KMeansModel.load(sc, "./KMeansModel")
sameModel: org.apache.spark.mllib.clustering.KMeansModel =
org.apache.spark.mllib.clustering.KMeansModel@68efc9a2
```

9. Output the loaded model:

```scala
scala> sameModel.toPMML
res3: String =
"<?xml version="1.0" encoding="UTF-8" standalone="yes"?>
<PMML version="4.2" xmlns="http://www.dmg.org/PMML-4_2">
    <Header description="k-means clustering">
        <Application name="Apache Spark MLlib" version="2.4.0"/>
        <Timestamp>2019-04-24T22:25:03</Timestamp>
    </Header>
    <DataDictionary numberOfFields="3">
        <DataField name="field_0" optype="continuous"
dataType="double"/>
        <DataField name="field_1" optype="continuous"
dataType="double"/>
        <DataField name="field_2" optype="continuous"
dataType="double"/>
    </DataDictionary>
    <ClusteringModel modelName="k-means" functionName="clustering"
modelClass="centerBased" numberOfClusters="2">
        <MiningSchema>
            <MiningField name="field_0" usageType="active"/>
```

Natural language processing for data analysis

Natural language processing (NLP) is the ability of a machine to analyze and understand human language. Human language has a very high amount of complexity, which makes parsing and understanding it difficult. There is a great deal of context in spoken and written language. Machines work well with precise rules that are within the confines of good context. With that said, it is still possible to gain an insight into text analysis using NLP techniques. An excellent example of this is Twitter sentiment analysis. Based on the contents of tweets, using NLP, it is possible to determine whether the sentiments of the people are generally positive or negative as a group. Another great example is the successful application of NLP techniques in analyzing customer reviews of a product or service.

The ML algorithms explored so far in this chapter make use of the variables in data that are for either numerical or categorical. NLP works with data that is nonnumerical and contains text information.

For running NLP in Spark, please refer to `https://github.com/JohnSnowLabs/spark-nlp` for detailed directions on how to run NLP. This is a separate project that is built on top of Spark and provides interesting ways to use NLP in Spark using Scala. For a quick start, please refer to `https://nlp.johnsnowlabs.com/quickstart.html`.

Algorithm selections

Each ML algorithm has its own strengths and weaknesses. Selecting an appropriate machine algorithm and tuning the model requires a fair amount of experience working with these algorithms, however, the following factors also play a significant role in applying these techniques effectively:

- **Asking the right question**: A great deal of effort is generally required in formulating the right question.
- **Understanding the business domain**: Having a good understanding of the relevant business domain and context is equally important to build good models.
- **Understanding data**: Ultimately, the data is used to train the model. If the data is not understood correctly or the data quality is poor, the built model is unlikely to be effective.

All of the preceding aspects outlined are somewhat interdependent and a mastery of all of these is a prerequisite to selecting the appropriate ML algorithm.

As a general rule, when the number of variables or features is relatively small and, intuitively, data can be easily split based on certain conditions, the decision tree could be the first choice. The model built by decision trees is easier to understand and comprehend. To make the model more effective, a random forest algorithm is the next choice, which can prevent the model from overfitting the training data.

When there is a large number of features in the dataset and the problem to be solved is a regression in nature, ridge and lasso regression algorithms work the best by reducing model complexity and preventing overfitting. The ridge algorithm reduces model complexity by coefficient shrinkage, whereas lasso reduces the number of features selected. There is another machine algorithm called **elastic net regression** that combines both of these aspects into model building. We have not discussed the elastic net regression algorithm in this chapter. The point to be noted here is that there are multiple ML algorithms that could be used to solve the same problem. It is often the case that one has to try and measure the effectiveness of each one of these. The outcome is generally dependent on the nature of the underlying data.

For problems that involve clustering, the k-means clustering algorithm works very well. Determining the value of k for the number of clusters requires a certain amount of experience and intuition about the data.

For analyzing human created text and voice data, the NLP algorithm is the only choice in the traditional ML space. We have not discussed deep learning algorithms in this chapter. Deep learning algorithms work quite differently from traditional ML algorithms. For such types of problems, deep learning algorithms are a great choice.

Summary

In this chapter, we learned about ML and some of the most popular ML algorithms. The primary goal of ML is to build an analytical model using historical data without much human intervention. ML algorithms can be divided into two categories, namely, supervised learning and unsupervised learning. The supervised learning algorithm relies on labeled data to build models, whereas unsupervised learning uses data that is not labeled. We looked at the k-means cluster analysis algorithm, which is an unsupervised ML algorithm. Of the supervised ML algorithms, we explored decision trees, random forests, and ridge/lasso regression. We also got an overview of using NLP for performing text data analysis.

In the next chapter, we will examine the processing of data in real time and perform data analysis as the data becomes available.

Section 3: Real-Time Data Analysis and Scalability

3

This section will introduce you to an emerging field of data analysis where you, analyze data in near real time using streaming technologies. In this section, you will learn about the concept of steam oriented processing and will be taken on a deeper dive into Spark steaming. You will also learn to analyse data from multiple dimensions, such as cost, reliability, and performance. This will provide you with a complete picture of how a practical real-world data analysis life cycle works and will prepare you for the future of data analysis.

This section will contain the following chapters:

- Chapter 8, *Near Real-Time Data Analysis Using Streaming*
- Chapter 9, *Working with Data at Scale*

8
Near Real-Time Data Analysis
Using Streaming

This chapter introduces another emerging and powerful technique in the field of data analysis—analyzing data in near real time using *Streaming technologies*. In the previous chapters, we looked at analyzing data that had already been created, using a technique known as **batch-oriented data processing**.

There are numerous cases where the value of data starts to diminish as the data starts to age. An excellent example of this is an online retailer that tracks customer interaction on its website. Offline batch-oriented analysis of this data to understand customer' behavior and preferences is certainly of great value to the retailer; however, a near real-time analysis of this data could have an even greater impact on the customer's experience. For example, a customer's experience could be made adaptive based on the current context in which the customer is interacting with the website.

To illustrate this with a more specific example, let's say that a customer has returned to the website after a long time and is searching for a specific category item in the catalog. To encourage the customer to purchase an item from that category, a discount offer could be made at that instant. In this case, time is of the essence because the customer may leave the website fairly quickly and not return for a considerable amount of time.

Overview of streaming

Stream processing is the act of continuously computing results as new data becomes available. A very simple example of this is computing the average of some numbers in a continuous fashion. To begin with, we start with the following information:

- Number of items = 0
- Current average = 0

As a new number comes in, we perform the following steps:

1. Compute a new total = Number of items x Current average + New number
2. Increment the number of items by one
3. Set the current average = New total / Number of items

As you can see, the continuous average computation algorithm is quite different from the batch-oriented algorithm. It is important to bear in mind the following facts when using this algorithm:

- The average value gets updated as new numbers become available
- The previously computed average value is reused to compute a new average

The following recipe using Scala code illustrates this:

1. Define a Scala function called `runningAverage` that will be used to provide an updated running average based on the previous running average and new items received, as follows:

```scala
scala> def runningAverage(prevAvgCount: Tuple2[Double, Long],
       newItems: Array[Int]): Tuple2[Double, Long] = {
     | val prevAverage = prevAvgCount._1
     | val prevItemCount = prevAvgCount._2
     | val newTotal = prevAverage * prevItemCount + newItems.sum
     | val newItemCount = prevItemCount + newItems.size
     | val newAverage = newTotal / newItemCount
     | Tuple2(newAverage, newItemCount)
     | }
runningAverage: (prevAvgCount: (Double, Long), newItems:
Array[Int])
               (Double, Long)
```

2. Using the following code, initialize the current average count as (0.0, 0), where 0.0 is the current average and 0 is the number of items:

```scala
scala> var currentAvgCount = Tuple2(0.0, 0L)

currentAvgCount: (Double, Long) = (0.0,0)
```

3. Compute the initial running average for three new items, (1, 2, 3), with the expected running average of (2.0, 3) using the following code:

```scala
scala> currentAvgCount = runningAverage(currentAvgCount,
                            Array(1,2,3))
currentAvgCount: (Double, Long) = (2.0,3)
```

4. Using the following code, update the current running average by adding 4 as a new item, creating an expected running average of (2.5, 4):

```scala
scala> currentAvgCount = runningAverage(currentAvgCount, Array(4))

currentAvgCount: (Double, Long) = (2.5,4)
```

5. Using the following code, repeat the preceding step for 5, creating the expected running average of (3.0, 5):

```scala
scala> currentAvgCount = runningAverage(currentAvgCount, Array(5))

currentAvgCount: (Double, Long) = (3.0,5)
```

6. Using the following code, repeat the preceding step for 6, creating the expected running average of (3.5, 6):

```scala
scala> currentAvgCount = runningAverage(currentAvgCount, Array(6))

currentAvgCount: (Double, Long) = (3.5,6)
```

7. Using the following code, repeat the preceding step for 7, creating the expected running average of (4.0, 7):

```scala
scala> currentAvgCount = runningAverage(currentAvgCount, Array(7))

currentAvgCount: (Double, Long) = (4.0,7)
```

The previously mentioned `runningAverage` method is a Scala function that is able to compute the new average given the following information:

- A tuple consisting of the previous average and previous item count
- An array of integers consisting of new items

If we compare the aforementioned algorithm to a simple average computation, we can observe some key differences:

1. Define a Scala function called `simpleAverage` that takes an array of items as input, computes the average of all items in the input, and returns this average value, as follows:

```scala
scala> def simpleAverage(items: Array[Int]): Double =
                        items.sum.toDouble / items.size

simpleAverage: (items: Array[Int])Double
```

2. Compute the average of (1, 2, 3) using the simpleAverage function, as follows:

```scala
scala> simpleAverage(Array(1, 2, 3))
res0: Double = 2.0
```

3. Repeat the average computation for arrays of different sizes, as follows:

```scala
scala> simpleAverage(Array(1, 2, 3, 4))
res1: Double = 2.5

scala> simpleAverage(Array(1, 2, 3, 4, 5))
res2: Double = 3.0

scala> simpleAverage(Array(1, 2, 3, 4, 5, 6))
res3: Double = 3.5

scala> simpleAverage(Array(1, 2, 3, 4, 5, 6, 7))
res4: Double = 4.0
```

Some of the key differences to be noted are as follows:

- The simpleAverage function requires all of the input items to be provided in order to compute the average. In contrast, runningAverage only needs previously computed results and new items to perform computations.
- simpleAverage does not hold and rely on any state information. On the other hand, runningAverage requires previously computed results to be preserved for the next iteration of computation.
- runningAverage can potentially operate on a very large number of items by working incrementally on small batches of items. For simpleAverage, all of the items have to be present at the time of computation, and its functionality becomes limited for a large number of items.

Although the preceding example is simple, it highlights the advantages of incremental and continuous data processing. The amount of recomputing that needs to be redone as a result of new data is generally significantly lower compared to reprocessing the entire dataset. For real-time or near real-time applications, fast response time is a necessity for success. Stream-oriented processing helps us to achieve that response time requirement. Stream processing does introduce some degree of complexity because some of the state information needs to be preserved, and the processing algorithm needs to be refactored and adapted to allow for continuous updates upon the arrival of new data.

The following diagram illustrates the general model of stream processing, where a **stream processor** is acting upon one or more observations at a time:

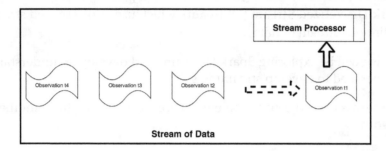

The follows facts should be noted regarding this generalized model:

- We can think of a stream as unbounded information written on tape. Each observation is recorded on the tape as it happens.
- **Observation t1** arrives before **Observation t2**, and so on. Observation t1 is written first, t2 next, and so on.
- The stream processor sees Observation t1 before t2, and so on. It computes the results based on t1 first, t2 next, and so on.

We can also imagine that streams of information are passing through the processor, and it is computing the results as and when this happens.

In reality, most streaming solutions use microbatches, where these accumulate certain amounts of information before handing it off to the processor. This is done to make the processing more efficient.

Spark Streaming overview

Spark Streaming is an extension of the core Spark API that enables scalable and fault-tolerant, stream-oriented processing of data. Spark provides the ability to stream data from multiple sources, with a number of key sources being the following:

- Apache Kafka
- Amazon Kinesis and S3
- TCP
- HDFS

Spark offers two flavors of streaming:

- Spark Structured Streaming that is built on top of the Spark SQL engine
- **Spark Discretized Stream** (**DStream**), which uses a discretized stream—that is, a continuous stream of data

In this section, we will be exploring Spark DStreams and develop an understanding of how this could be leveraged to develop streaming solutions.

Let's start with a classic word count problem, where we are trying to count the frequency of each distinct word.

Word count using pure Scala

Generating a word count is a classic problem that is widely used to demonstrate some key concepts related to solving problems involving a massive dataset. At a high level, the objective of the word count problem is to count the frequencies of each distinct word in a document.

First, let's look at solving this problem in Scala without using Spark. For the sake of simplicity, we will treat the same word with different cases as distinct words. Try using the following Scala REPL as follows:

1. Create some sample text data using the following code:

```
scala> val text = "This is a sample for testing word count example
        It should count the frequency of each distinct word"

text: String = This is a sample for testing word count example It
               should count the frequency of each distinct word
```

2. Split the data using white space as the delimiter, as follows. This provides us with an array of strings:

```
scala> val words = text.split("\\s+") // split on white spaces

words: Array[String] = Array(This, is, a, sample, for, testing,
                       word, count, example, It, should,
                       count, the, frequency, of, each,
                       distinct, word)
```

3. Group these same words together by using the `groupBy` method of the Scala array, as follows. We get a mapping from word to array, where the word is repeated as many times as it occurs in the data:

```
scala> val grouped = words.groupBy(w => w) // group same words
together

grouped: scala.collection.immutable.Map[String,Array[String]] =
Map(for -> Array(for), count -> Array(count, count), is ->
Array(is), This -> Array(This), a -> Array(a), each -> Array(each),
testing -> Array(testing), should -> Array(should), distinct ->
Array(distinct), sample -> Array(sample), It -> Array(It),
frequency -> Array(frequency), example -> Array(example), word ->
Array(word, word), of -> Array(of), the -> Array(the))
```

4. Count the number of words in each array by applying the `mapValues` method, as follows. This provides us with another map, where the key represents a distinct word and the value represents the number of occurrences of that word:

```
scala> val wordCount = grouped.mapValues(_.size) // count the
number of elements for each key

wordCount: scala.collection.immutable.Map[String,Int] = Map(for ->
1, count -> 2, is -> 1, This -> 1, a -> 1, each -> 1, testing -> 1,
should -> 1, distinct -> 1, sample -> 1, It -> 1, frequency -> 1,
example -> 1, word -> 2, of -> 1, the -> 1)
```

There are multiple ways to solve the word count problem in Scala. The preceding example is one such way. We performed the following actions:

1. Split the text into an array of words by splitting it with white spaces.
2. Grouped the same words together into a map whose key is the word and whose values are an array of the word repeated as the occurrence of that word.
3. Next, we mapped the values of grouped results into the count of the number of elements. This gave us a map of the distinct word to the associated count.

Word count using Scala and Spark

Next, let's look at solving the same word count problem in Spark. Try going through the following steps:

1. Start a Spark shell session in your Terminal, as follows:

```
$ spark-shell
```

2. Define the sample test data. We will use the same text as the previous example, as shown in the following code:

```
scala> val text = "This is a sample for testing word count example
                   It should count the frequency of each distinct
                   word"

text: String = This is a sample for testing word count example It
               should count the frequency of each distinct word
```

3. Convert the sample data into a Spark **resilient distributed dataset (RDD)** using `SparkContext`, given as the variable `sc` in the session, as shown in the following code:

```
scala> val rdd = sc.parallelize(Seq(text)) // convert to an RDD

rdd: org.apache.spark.rdd.RDD[String] = ParallelCollectionRDD[8] at
parallelize at <console>:26
```

4. Split the data using white spaces as the delimiter and map each word to a pair (word, 1) and sum the value according to distinct word, as follows:

```
// flat map of splits, word to count 1 and reduce by adding counts
// of a word
scala> val wc = rdd.flatMap(_.split("\\s+")).map(w => (w,
                 1)).reduceByKey(_+_)

wc: org.apache.spark.rdd.RDD[(String, Int)] = ShuffledRDD[11] at
reduceByKey at <console>:25
```

5. Finally, collect the results to force Spark to execute the recipe we built earlier, as follows:

```
scala> wc.collect
res0: Array[(String, Int)] = Array((testing,1), (a,1), (each,1),
                            (for,1), (the,1), (example,1), (is,1),
                            (word,2), (sample,1), (should,1),
                            (It,1), (frequency,1), (distinct,1),
                            (This,1), (of,1), (count,2))
```

As we can see in the previous steps, we are able to produce the same results using Spark; however, the recipe has the following important differences compared to a pure Scala solution:

- We had to create an RDD from the input text. At first, this might seem an inconvenient and unnecessary step; however, RDD is the feature of Spark that provides the foundation for distributed computing and performing data processing at scale.
- We used the `flatMap` API of RDD to split each word as an individual item.
- Next, we mapped each word as a tuple of the word and 1 as the count.
- We used the `reduceByKey` API of RDD to sum the count of each word. This gave us a new RDD that had the desired results of the words and their associated counts.
- We performed a collect operation on RDD to gather the results. Please note that Spark transformations, such as `flatMap`, `map`, and `reduceByKey`, are all lazily evaluated. An action such as collect materializes the results.

Word count using Scala and Spark Streaming

As a next step, we will look at solving this problem in a streaming fashion using Spark DStreams. We will use the **Transmission Control Protocol (TCP)** as a means of sending data to the stream processor using a tool called `netcat` (or `nc`). The `nc` tool is a simple but powerful utility tool available on most Linux distributions and macOS. The recipe here uses macOS, as follows:

1. Let's first start the `nc` server process on one Terminal, as follows:

```
$ nc -kl 12345 # 12345 is the port number server listens
```

2. On another Terminal, start the nc client that connects to this server:

```
$ nc localhost 12345 # client connects to server on localhost at
port 12345
```

3. At this point, the client has established a connection. Now, enter some text on the server screen and hit *Enter*. This text should be displayed back on the client screen as well, showing text similar to the following:

```
$ nc -kl 12345 # Server
Here is some text
Here is some more text

$ nc localhost 12345 # Client
Here is some text
Here is some more text
```

This test is to make sure that the nc setup is working correctly. For the Spark DStreams exercise, we just need the nc server process to be running since the Spark DStreams process will act as a client process. You can kill the client process now by pressing *Control + C* on that Terminal. Please make sure that the nc server process keeps running. Here is an overview of data flow between the three participants:

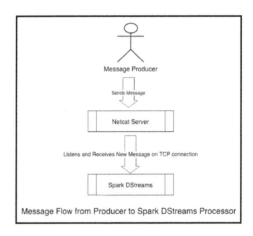

Message Flow from Producer to Spark DStreams Processor

The preceding diagram illustrates the following:

- The **Message Producer**—we will act as a message producer by typing in a message on the netcat server Terminal
- The **Netcat Server** will receive the typed message
- The **Spark DStreams** will receive the message from the netcat server almost instantly using TCP

Now, we will set up a Spark Streaming context by going through the following steps:

1. Start the Spark shell, as follows:

   ```
   $ spark-shell
   ```

2. Using the following code, stop the current Spark session, because the default session does not support streaming:

   ```
   scala> spark.stop() // stop the existing Spark session
   ```

3. Import the `org.apache.spark` package using the following; this is needed for setting the Spark configuration for Spark Streaming:

   ```
   scala> import org.apache.spark._import org.apache.spark._
   ```

4. Import the `org.apache.spark.streaming` package using the following; this is needed for setting up the Spark Streaming session:

   ```
   scala> import org.apache.spark.streaming.
                    _importorg.apache.spark.streaming._
   ```

5. Create a new Spark configuration using the following code, making sure that at least two cores are reserved:

   ```
   scala> val conf = new
   SparkConf().setMaster("local[2]").setAppName("HandsOnSparkStreaming
   ") // least two cores are required to prevent a starvation scenario
   conf: org.apache.spark.SparkConf =
   org.apache.spark.SparkConf@3031d9e9
   ```

6. Create a Spark `StreamingContext` using the preceding Spark configuration, as follows:

   ```
   scala> val ssc = new StreamingContext(conf, Seconds(5)) // 5
   seconds: the time interval at which streaming data will be divided
   into batches

   ssc: org.apache.spark.streaming.StreamingContext =
   org.apache.spark.streaming.StreamingContext@40874f54
   ```

7. Set up a `.checkpoint` directory as follows:

   ```
   scala> ssc.checkpoint(".") // use current directory for checkpoint
   ```

We have already done the following:

- Stopped the existing Spark session because we needed a `StreamingContext` instead.
- Set up a `StreamingContext` with two cores.
- Set up the current directory as a checkpoint. This is needed for stateful stream processing.

Next, let's define the `RunningUpdate` Scala `case` object and test it by going through the following steps:

1. Define `RunningUpdate`, as follows:

   ```
   scala> // Define a Serializeable running update function

   scala> case object RunningUpdate { // this makes the object
   Serializeable
        | val updateCount = (newValues: Seq[Int], runningCount:
          Option[Int]) => {
        | val newCount = runningCount.getOrElse(0) + newValues.sum
        | Some(newCount): Option[Int]
        | }
        | }

   defined object RunningUpdate
   ```

The `RunningUpdate` object has the following properties:

- It is a Scala `case` object, which makes this serializable. This is needed for it to work with Spark Streaming because of its distributed nature.
- It holds a method, `updateCount`, which takes a sequence of new values and the running count. This method returns the new running count.

2. Test the `RunningUpdate`, as follows:

   ```
   scala> RunningUpdate.updateCount(Seq(1, 2, 3), Some(10)) // test
   update: expected 1+2+3+10 = 16

   res2: Option[Int] = Some(16)
   ```

The Spark Streaming framework will invoke this `updateCount` method automatically, as we will see:

1. Set up a socket text stream on the localhost at port `12345` as follows. Please note the our `nc` server is running on the localhost at this port. By doing this, we can start receiving messages from the `nc` server in a streaming fashion:

```
scala> val lines = ssc.socketTextStream("localhost", 12345) //
12345 is the netcat server port

lines:
org.apache.spark.streaming.dstream.ReceiverInputDStream[String] =
org.apache.spark.streaming.dstream.SocketInputDStream@4bb05a4e
```

2. Split each line in the stream into words, as follows:

```
scala> val words = lines.flatMap(_.split("\\s+"))
words: org.apache.spark.streaming.dstream.DStream[String] =
org.apache.spark.streaming.dstream.FlatMappedDStream@60877629
```

3. Map each word as a `(word, 1)` key–value pair, as follows:

```
scala> val pairs = words.map(word => (word, 1))

pairs: org.apache.spark.streaming.dstream.DStream[(String, Int)] =
org.apache.spark.streaming.dstream.MappedDStream@531f6879
```

4. Apply the `updateCount` function, `RunningUpdate`, in a stateful way, as follows:

```
scala> val runningCounts =
pairs.updateStateByKey[Int](RunningUpdate.updateCount) // apply
running update on stream

runningCounts: org.apache.spark.streaming.dstream.DStream[(String,
Int)] = org.apache.spark.streaming.dstream.StateDStream@4819a11f
```

5. Print out sample running counts, as follows:

```
scala> runningCounts.print()
```

6. So far, we have built the recipe for streaming. The stream processing has not started as yet. Start stream processing now using the following code:

```
scala> ssc.start() // Start the computation
```

7. Set up a reasonable timeout to receive the control back into the Spark shell, as follows:

```
scala> ssc.awaitTerminationOrTimeout(1000*60*5) // Wait for 5
minutes before timing out
```

We have done the following in the preceding code example:

- Connected to our `nc` server running on the localhost at port `12345` to receive messages as single lines of text.
- Applied a recipe similar to Spark on each line of text by using `flatMap` and `map` APIs.
- Applied a running update state to each pair by running them through the `RunningUpdate.updateCount` function.
- Printed running counts.
- Started the Spark Streaming process. Work starts to get performed only after the invocation of this method.

On the netcat screen, input some text messages introducing some intermediate delays between each message typing. The following code is an example of this:

```
$ nc -kl 12345
hello world
hello
hello
world
how are you
hello again
```

On your Spark shell, you should see the counts being updated as this happens, along with some debug information outputted by Spark Stream. The following code is an example based on the input provided using the preceding netcat example:

```
...
(hello,1)
(world,1)
...
(hello,2)
(world,1)
...
(hello,3)
(world,2)
...
(are,1)
(how,1)
```

```
(hello,4)
(again,1)
(world,2)
(you,1)
```

Run the following code to gracefully stop the Spark Streaming process and quit Spark shell:

```
scala> ssc.stop()

scala> :quit
```

You can also terminate the netcat server process if required by pressing *Control + C* on the Terminal.

You should now see that Spark Streaming provides a powerful framework for performing continuous computations as new data arrives. It also provides new UI-based tools to look at stream processing details and the performance statistics of the job. This can be done by going to the URL, `http://localhost:4040`, in your web browser. The first set of interesting information is found under the **Jobs** tab, as shown in the following screenshot:

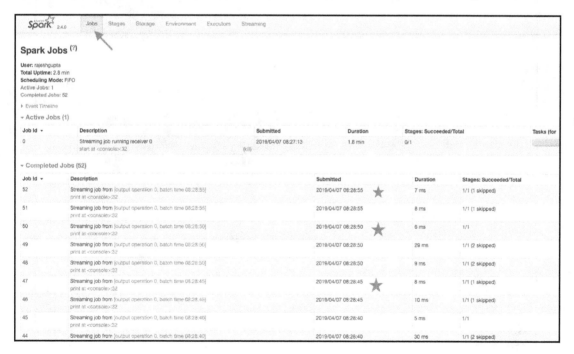

The **Jobs** tab shows the overall progress of the job. You can see that the work is being completed at intervals of five seconds, since we defined our batch window as a five-second interval. Depending on the use case, an appropriate value for the interval can be chosen. In an application where the response time is critical, a lower value can be selected. A higher value provides more efficiency, since processing happens in bigger batches; however, this comes at the cost of a slower response time.

Another interesting tab to look at in this UI is the **Streaming** tab, as shown in the following screenshot:

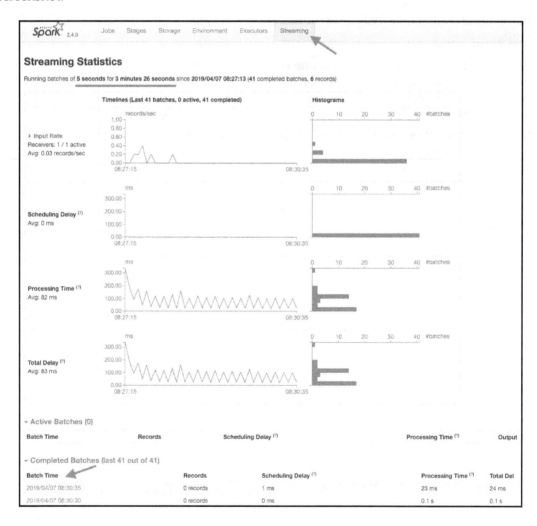

The **Streaming** tab provides performance statistics from several different perspectives. This is a very useful tool for streaming jobs that run in production because it provides a fairly good insight into any potential delays with stream processing. One of the main reasons for moving toward stream-oriented processing is that it performs computations as new data arrives and provides insight based on the new information. Any unexpected delays in processing data would be undesirable, and this specific dashboard can act as a good tool to determine whether delays are occurring.

Deep dive into the Spark Streaming solution

Let's look at how the Spark Streaming solution was used to solve the word count problem. There are a few important pieces of information that we need to take a look at in order to understand this solution.

The first important piece is the `RunningUpdate` object, which is a Scala case object. One important feature of the Scala `case` object and objects created from the Scala `case` class is that the objects are serializable. This object holds a running update function that is assigned to the `updateCount` Scala value, as follows:

```
case object RunningUpdate { // this makes the object serializable
  val updateCount = (newValues: Seq[Int], runningCount: Option[Int]) => {
    val newCount = runningCount.getOrElse(0) + newValues.sum
    Some(newCount): Option[Int]
  }
}
```

This object needs to be serializable because Spark is a distributed system, and it ships the code to workers. The code shipped must be serializable and a worker will receive this code. The worker deserializes the code received and applies this code to a piece of data it is working on.

The second important piece of information is the following code that uses the `updateStateByKey` API of Spark Streaming:

```
val runningCounts = pairs.updateStateByKey[Int](RunningUpdate.updateCount)
```

We are passing the `RunningUpdate` object's `updateCount` function as a parameter to the `updateStateByKey` method on the pairs stream. We can envision the pairs stream as a mapping from the word to the current count, as follows:

- The word is the key
- The current count is the value

It is also not the signature of `updateCount`. The first parameter is a sequence of integers, where the second parameter is an optional parameter. The following list goes into more detail:

- `newValues`: A sequence of integers containing newly observed counts for a specific word.
- `runningCount`: The current running count for the same word. This value is `none` at the start of the stream and when a word is observed for the first time by the stream. Its value is `Some(newcount)` if the word has been seen earlier.

The final import part is the `flatMap` and `map` APIs, as shown in the following code:

```
val ssc = new StreamingContext(conf, Seconds(5))
...
val lines = ssc.socketTextStream("localhost", 12345) // 12345 is the netcat
server port
val words = lines.flatMap(_.split("\\s+"))
val pairs = words.map(word => (word, 1))
val runningCounts = pairs.updateStateByKey[Int](RunningUpdate.updateCount)
// apply running update on stream
runningCounts.print()
ssc.start() // Start the computation
```

Even though `flatMap` and `map` seem to be working like standard Spark APIs, they are working on Spark DStreams. They have the following characteristics:

- `lines` represent a minibatch of data received within the 5 second window
- `flapMap` and `map` are applied to this minibatch
- The same is true for `updateStateByKey`

The stream is started only when `ssc.start()` is executed, and the following piece of the recipe is repeated over and over again for each batch window until the stream is stopped:

```
val words = lines.flatMap(_.split("\\s+"))
val pairs = words.map(word => (word, 1))
val runningCounts = pairs.updateStateByKey[Int](RunningUpdate.updateCount)
// apply running update on stream
runningCounts.print()
```

We can also rewrite the preceding code in the following way with identical results:

```
ssc.socketTextStream("localhost", 12345).
  flatMap(_.split("\\s+")).
  map(word => (word, 1)).
  updateStateByKey[Int](RunningUpdate.updateCount).
  print()
```

Please note that `socketTextStream` is called only once; however, the following are called for each minibatch, in this order:

- `flatMap`
- `map`
- `updateStateByKey`
- `print`

Since this is a stateful stream, we also need checkpointing to persist states, as follows:

```
ssc.checkpoint(".") // use current directory for checkpoint
```

For more information on the checkpoint, please refer to the Apache Spark documentation for Spark DStreams at `https://spark.apache.org/documentation.html`.

Let's now put this all together in plain Scala to understand the flow by going through the following steps. We will use the same inputs as we provided to the netcat server:

1. Define an `updateCount` method as follows:

```
scala> val updateCount = (newValues: Seq[Int], runningCount:
Option[Int]) => {
    | val newCount = runningCount.getOrElse(0) + newValues.sum
    | Some(newCount): Option[Int]
    | }
updateCount: (Seq[Int], Option[Int]) => Option[Int] =
$$Lambda$3671/605827193@2fdd23c1
```

2. Create `inputs` as arrays of texts, as follows:

```scala
scala> val inputs = Array("hello world", "hello", "hello", "world",
"how are you", "hello again")

inputs: Array[String] = Array(hello world, hello, hello, world, how
are you, hello again)
```

3. Import Scala's `mutable.Map`, as follows. This is needed because we will need a map that cannot be updated:

```scala
scala> import collection.mutable.Map
import collection.mutable.Map
```

4. Create a `mutable.Map` as follows:

```scala
scala> val wordToCount = Map[String, Int]()
wordToCount: scala.collection.mutable.Map[String,Int] = Map()
```

5. For each element in the inputs, process and apply the update as follows:

```scala
scala> inputs.foreach(i => {val words = i.split("\\s+").groupBy(w
=> w).mapValues(vs => vs.map(v => 1).toSeq).foreach(
     | wc => {
     | val word = wc._1
     | val newValues = wc._2
     | val runningCount = if (wordToCount.contains(word))
Some(wordToCount(word)) else None
     | val updatedCount = updateCount(newValues, runningCount)
     | wordToCount(word) = updatedCount.get
     | }); println(wordToCount);
     | })
Map(world -> 1, hello -> 1)
Map(world -> 1, hello -> 2)
Map(world -> 1, hello -> 3)
Map(world -> 2, hello -> 3)
Map(you -> 1, how -> 1, world -> 2, are -> 1, hello -> 3)
Map(again -> 1, you -> 1, how -> 1, world -> 2, are -> 1, hello ->
4)
```

We defined a `mutable.Map` that holds the running count for each unique word. To begin with, the `Map` is empty. As we iterate through each input record, the counts are updated for each observed word. When we use Spark Streaming, this all happens under the hood, and the streaming framework repeats the loop with each new batch of data. It also keeps track of the state in a similar way to the `mutable.Map` and `wordToCount` method that we described previously.

The preceding illustration provides some insight into what Spark Streaming does to process this data continuously. One of the biggest advantages of using a framework such as Spark is that it also provides stream processing with fault tolerance. Fault tolerance semantics can become fairly complex in the case of stateful processing, such as the example we just explored. Spark provides certain guarantees of how this works. Please refer to the Apache Spark Streaming documentation for more details on this at `https://spark.apache.org/documentation.html`.

Streaming a k-means clustering algorithm using Spark

The k-means algorithm is an unsupervised **machine learning** (**ML**) clustering algorithm. The objective of this algorithm is to build k centers around which data points are centered, thereby forming k clusters. The most common implementation of this algorithm is generally done using batch-oriented processing. Streaming-based clustering algorithms are also available for this, with the following properties:

- The k clusters are built using initial data
- As new data arrives in minibatches, existing k clusters are updated to compute new k clusters
- It also possible to control the decay or decrease in the significance of older data

At a high level, the preceding steps are quite similar to the word count problem that we solved using the streaming solution. The goal of the k-means algorithm is to partition the data into k clusters. If the data is static and does not change, then k clusters can be formed and will also remain static. On the other hand, if new data is arriving, new k clusters need to be computed, and these will change based on the combined properties of old and new data. A streaming-based k-means algorithm works on a minibatch of new data.

In Spark Streaming and Spark MLlib , the clusters can be computed dynamically as the new data arrives as part of a data stream. The Spark API is defined here at `https://spark.apache.org/docs/2.4.1/api/java/index.html?org/apache/spark/mllib/clustering/StreamingKMeans.html`.

To initialize a `StreamingKMeans` object, the following needs to be done using the builder pattern:

- Instantiate the `StreamingKMeans` object—new `StreamingKMeans()`
- Set the number of clusters as `setK(k)`
- Set the decay factor using `setDecayFactor(decayFactor)`
- Set initial random centers using `setRandomCenters(dimension, weight, seed)`
- We finally get a model that can be trained in a streaming fashion and can be used to make predictions

Let's set this up in a Spark shell by going through the following steps:

1. Start a Spark shell on the Terminal, as follows:

   ```
   $ spark-shell
   ```

2. Using the following code, stop the current Spark session because it does not support Streaming. We need to explicitly create a Spark Streaming session:

   ```
   scala> spark.stop() // stop current Spark session
   ```

3. Import the Spark package using the following code. This is needed to configure Spark:

   ```
   scala> import org.apache.spark._
   import org.apache.spark._
   ```

4. Using the following code, import the Spark Streaming package, which is needed to set up the Spark Streaming session:

   ```
   scala> import org.apache.spark.streaming._
   import org.apache.spark.streaming._
   ```

5. Import the `StreamingKMeans` class from the Spark MLlib package, as follows:

   ```
   scala> import org.apache.spark.mllib.clustering.StreamingKMeans
   import org.apache.spark.mllib.clustering.StreamingKMeans
   ```

6. Create a new Spark configuration as follows, making sure that at least two cores are allocated for processing:

```
scala> val conf = new
SparkConf().setMaster("local[2]").setAppName("HandsOnKMeanStreaming
")

conf: org.apache.spark.SparkConf =
org.apache.spark.SparkConf@1483c738
```

7. Create a Spark Streaming context as follows:

```
scala> val ssc = new StreamingContext(conf, Seconds(30))
ssc: org.apache.spark.streaming.StreamingContext =
        org.apache.spark.streaming.StreamingContext@7a5a26b7
```

8. Initialize the `StreamingKmeans` object:

```
scala> val model = new StreamingKMeans().
 | setK(4). // number of clusters
 | setDecayFactor(1.0). // decay
 | setRandomCenters(3, 0.0) // 3 dimensions and 0 weight
model: org.apache.spark.mllib.clustering.StreamingKMeans =
        org.apache.spark.mllib.clustering.StreamingKMeans@2e1b1d22
```

The preceding code is quite similar to the one used for the word count, with the following important properties:

- We set the Spark Streaming context, `ssc`, in exactly the same way as the Streaming word count problem
- We initialized a Spark Streaming k-means model with the desired set of parameters

The next steps are about creating a DStream that provides the input data and trains the model using new batches of data coming on Spark DStreams, and then finally starts the Streaming context. Look at the following code:

```
...
model.trainOn(trainingData) // traningData is a Spark DStreams with data
coming in mini batches
...
ssc.start()
...
```

To see this in action, let's start with some data preparation. Create a new temporary directory in a suitable location for training and test data. One good choice for such data is the /tmp directory on Linux and macOS, as shown in the following code:

```
$ mkdir /tmp/k-means-train-data
$ mkdir /tmp/k-means-test-data
```

We will work with three-dimensional data to explore and understand this. The format for three-dimensional training data is as follows:

```
[n1, n2, n3]
```

The following is an example of training data:

```
[1000, 0, 0]
[1001, 0, 0]
[1002, 0, 0]
[1003, 0, 0]
[1004, 0, 0]
[1005, 0, 0]
[1006, 0, 0]
[1007, 0, 0]
[1008, 0, 0]
[1009, 0, 0]
[0, 1000, 0]
[0, 1001, 0]
[0, 1002, 0]
[0, 1003, 0]
[0, 1004, 0]
[0, 1005, 0]
[0, 1006, 0]
[0, 1007, 0]
[0, 1008, 0]
[0, 1009, 0]
[0, 0, 1000]
[0, 0, 1001]
[0, 0, 1002]
[0, 0, 1003]
[0, 0, 1004]
[0, 0, 1005]
[0, 0, 1006]
[0, 0, 1007]
[0, 0, 1008]
[0, 0, 1009]
```

The format of test data is as follows:

```
(num, [n1, n2, n3])
```

The following is an example of test data:

```
(1, [1002, 0, 0])
(1, [1001, 0, 0])
(2, [0, 1002, 0])
(2, [0, 1001, 0])
(3, [0, 0, 1002])
(3, [0, 0, 1001])
```

To start exploring this algorithm, start a Spark shell and go through the following steps:

1. Start a Spark shell using the following command:

   ```
   $ spark-shell
   ```

2. Stop the current Spark session using the following command:

   ```
   spark.stop() // stop current Spark session
   ```

3. Import the necessary packages and classes as follows:

   ```
   import org.apache.spark._
   import org.apache.spark.streaming._
   import org.apache.spark.mllib.clustering.StreamingKMeans
   ```

4. Set up the Spark configuration and streaming context as follows:

   ```
   val conf = new
   SparkConf().setMaster("local[2]").setAppName("HandsOnKMeanStreaming
   ")
   val ssc = new StreamingContext(conf, Seconds(10))
   ```

5. Create a `StreamingKMeans` method by initializing the parameters. The following code is the model:

   ```
   val model = new StreamingKMeans().
    setK(4). // number of clusters is 4
    setDecayFactor(1.0). // decay factor (the forgetfulness of the
   previous centroids)
    setRandomCenters(3, 0.0) // 3 dimensions and 0 weight
   ```

6. Set up the training data stream as follows:

```
import org.apache.spark.mllib.linalg.Vectors
val trainingData = ssc.textFileStream("file:/tmp/k-means-train-
data").map(Vectors.parse).cache()
trainingData.print() // to output training data for debug purpose
```

7. Set up the test data stream as follows:

```
import org.apache.spark.mllib.regression.LabeledPoint
val testData = ssc.textFileStream("file:/tmp/k-means-test-
data").map(LabeledPoint.parse)
```

8. Train the model on the data from the training data stream as follows:

```
model.trainOn(trainingData)
```

9. Use the model to predict how to use the data from the test data stream, as follows:

```
model.predictOnValues(testData.map(lp => (lp.label,
                                      lp.features))).print()
```

10. Start the streaming processing, as follows:

```
ssc.start()
```

11. Set a timeout for streaming processing to run before getting control back in the shell, as follows:

```
ssc.awaitTerminationOrTimeout(1000*60*3) // Wait for the
computation to terminate (3 minutes)
```

12. Remember to stop the streaming context cleanly and exit the Spark shell when it is finished, as follows:

```
scala> ssc.stop()

scala> :quit
```

Go through the following steps to see k-means streaming-based clustering in action:

1. Copy a text file containing some training data in the previously specified training data format to the /tmp/k-means-train-data directory.
2. Copy a text file containing some test data in the previously specified test data format to the /tmp/k-means-test-data directory.

3. Repeat the preceding steps with new training and test data copied to the respective directories. You should now see the clusters being updated continuously.

In your Spark shell, you can also see some additional details pertaining to the latest model, as shown in the following code:

```
. . .
scala> model.latestModel.clusterCenters
res15: Array[org.apache.spark.mllib.linalg.Vector] =
Array([0.0,1004.5,0.0],
[502.2500000000017,3.3333333333333336E-15,502.2500000000017],
[502.249999999995,-1.0E-14,502.249999999995])
. . .
scala> model.latestModel.clusterWeights
res16: Array[Double] = Array(20.0, 30.0, 10.0)
```

We initially started with random centers and zero weights. As the training data passes through the streaming k-means `trainOn` API, it recomputes new centers and weights. Using the decay factor, the significance of aged data can also be controlled.

Streaming linear regression using Spark

Using Spark Streaming, it is possible to update the parameters of the linear model online. In many ways, Spark Streaming's linear regression solution works very similarly to the k-means streaming solution.

We will be using the `StreamingLinearRegressionWithSGD` class that is provided as part of Spark MLlib. To initialize a `StreamingLinearRegressionWithSGD` object, the following needs to be done:

1. Instantiate the `StreamingLinearRegressionWithSGD` object using the `new StreamingLinearRegressionWithSGD()` method
2. Set the number of initial weights
3. We should get a model that can be trained in a streaming fashion and can be used to make predictions

Let's explore this solution in a Spark shell by going through the following steps:

1. Start a Spark shell in your Terminal as follows:

   ```
   $ spark-shell
   ```

2. Stop the current Spark session using the following code, since we need to explicitly create a streaming session:

   ```
   scala> spark.stop() // stop current Spark session
   ```

3. Using the following code, import the spark package, which is required to configure Spark:

   ```
   scala> import org.apache.spark._
   import org.apache.spark._
   ```

4. Use the following code to import the Spark Streaming package, which is needed for Spark Streaming:

   ```
   scala> import org.apache.spark.streaming._
                 import org.apache.spark.streaming._
   ```

5. Use the following code to import the Vectors from Spark MLlib's linear algebra package:

   ```
   scala> import org.apache.spark.mllib.linalg.Vectors
   import org.apache.spark.mllib.linalg.Vectors
   ```

6. Use the following code to import LabelPoint from Spark MLlib's regression package:

   ```
   scala> import org.apache.spark.mllib.regression.LabeledPoint
   import org.apache.spark.mllib.regression.LabeledPoint
   ```

7. Import the StreamingLinearRegressionWithSGD class from Spark MLlib's regression package using the following code:

   ```
   scala> import
   org.apache.spark.mllib.regression.StreamingLinearRegressionWithSGD
   import
   org.apache.spark.mllib.regression.StreamingLinearRegressionWithSGD
   ```

8. Using the following code, set up Spark's configuration, making sure that there are at least two cores allocated for processing:

```scala
scala> val conf = new
SparkConf().setMaster("local[2]").setAppName("HandsOnStreamingLinea
rReg")
conf: org.apache.spark.SparkConf =
org.apache.spark.SparkConf@403b7be3
```

9. Set up the Spark Streaming context using the following code:

```scala
scala> val ssc = new StreamingContext(conf, Seconds(10))
ssc: org.apache.spark.streaming.StreamingContext =
org.apache.spark.streaming.StreamingContext@72bef795

scala> val numFeatures = 3
numFeatures: Int = 3

scala> val model = new StreamingLinearRegressionWithSGD().
 | setInitialWeights(Vectors.zeros(numFeatures))
model:
org.apache.spark.mllib.regression.StreamingLinearRegressionWithSGD
=
org.apache.spark.mllib.regression.StreamingLinearRegressionWithSGD@
5db6083b
```

10. Set the number of features to 3 (as we will be working with three-dimensional data) using the following code:

```scala
scala> val numFeatures = 3
numFeatures: Int = 3
```

11. Initialize the model, as follows:

```scala
scala> val model = new StreamingLinearRegressionWithSGD().
 | setInitialWeights(Vectors.zeros(numFeatures))
model:
org.apache.spark.mllib.regression.StreamingLinearRegressionWithSGD
=
org.apache.spark.mllib.regression.StreamingLinearRegressionWithSGD@
5db6083b
```

Just like the k-means streaming solution, the next step is to create a DStreams that provides the input data and trains the model using new batches of data coming on Spark DStreams. Then, we can finally start the Streaming context. Let's set this up by going through the following steps:

1. Set up the training data stream as follows:

```
val trainingData = ssc.textFileStream("file:/tmp/lin-reg-train-
                      data").map(Vectors.parse).cache()
trainingData.print() // to output training data for debug purpose
```

2. Set up the test data stream as follows:

```
val testData = ssc.textFileStream("file:/tmp/lin-reg-test-
                   data").map(LabeledPoint.parse)
```

3. Train the model on data from the training data stream, as follows:

```
model.trainOn(trainingData)
```

4. Use the model to predict using data from the test data stream, as follows:

```
model.predictOnValues(testData.map(lp => (lp.label,
lp.features))).print()
```

5. Start the processing of streaming using the following command:

```
ssc.start()
```

6. Set a timeout for the streaming processing to run before getting control back in the shell, as follows:

```
ssc.awaitTerminationOrTimeout(1000*60*3) // Wait for the
computation to terminate (3 minutes)
```

7. Remember to stop the streaming context cleanly and exit the Spark shell when finished, as follows:

```
scala> ssc.stop()

scala> :quit
```

To see this in action, let's start with some data preparation. Create a new temporary directory in a suitable location for training and test data. One good choice for this location is the /tmp directory on Linux and macOS, as shown in the following code:

```
$ mkdir /tmp/lin-reg-train-data
$ mkdir /tmp/lin-reg-test-data
```

We will work with three-dimensional data to explore and understand this. The format for three-dimensional training data is as follows:

```
(num, [n1, n2, n3])
```

Since linear regression is a supervised ML algorithm, we need to provide labeled data for training. The following list shows the labels that are used in detail:

- [n1, n2, n3] represents a data point of three features
- num represents the associated label or known observation

Summary

In this chapter, we learned how to process data in near real time using a streaming-based approach. Streaming processing is quite different from traditional batch-oriented processing. Through a classic word count example, we explored how streaming-oriented processing could be applied to such problems to get near real-time updates. The streaming algorithm is quite different from the classic solution to this problem, and introduces a few complex concepts, such as state management. For all the added complexity in the streaming solution, it is generally worth employing because of the significantly improved response time in gaining real-time details of the data that is being monitored.

We also looked at how to make use of the streaming-oriented approach for ML. In the next chapter, we will look at scalability concerns.

Working with Data at Scale

Data is being produced at an accelerated pace with advancements in technology. The widespread usage and adoption of the **Internet of Things (IoT)** is a great example of this. These specifically purposed IoT devices are tens of billions in number and are growing rapidly. Many of these devices, using their sensors, continually produce observations as data. Even though the data might be small as a unit, combined together it becomes humongous. IoT is just one example of how much and how fast the data is being created.

This kind of data is sometimes referred to as **big data** that is too big to fit on a single machine for storage and computing purposes. Big data has three important properties:

- **Variety**: Data in different formats and structures
- **Velocity**: New data arriving at a fast rate
- **Volume**: Huge overall data size

In the prior chapters, we learned how to deal with a variety of data formats in Scala. For example, we explored Scala libraries for processing data in CSV, XML, and JSON formats.

We also explored various data processing techniques that fall into two categories:

- Batch-oriented
- Stream-oriented

Stream-oriented processing helps us continually perform computations as the data arrives. On the other hand, batch-oriented processing provides more efficiency at the cost of higher latency in the availability of computed results. Stream-oriented processing is a powerful tool for dealing with data velocity problems. In the example of data created by IoT devices, the value of data generally starts to decrease as the data ages, and low latency data processing becomes one of the dominant factors in the design of such systems.

Dealing with large volumes of data is a different kind of challenge. The data needs to be organized and stored at a large scale. It also needs to be processed for various purposes within the desired time periods to extract meaningful business values from it. In this chapter, we will focus on the volume aspect of data.

In this chapter, we will cover the following topics:

- Working with data at scale
- Cost considerations
- Reliability considerations

Working with data at scale

Working with data at scale and handling large data volumes significantly changes data analysis and processing. To get an intuition for the problems with data at scale, let's look at a simple problem of computing the median value of numbers. The median is the mid-point that splits the data into two parts. Use the following numbers as an example:

```
8 1 2 7 9 0 5
```

We will first sort the numbers in ascending order:

```
0 1 2 5 7 8 9
```

The median value is 5, because it splits the data into two halves, where half of the values are below five and the another half are above five.

Now, let's imagine that the count of these numbers was of the order of billions. Let's explore a solution to this problem in Scala REPL. Traditionally, we would need to do the following steps to compute the median value:

1. Load the data into memory on a single computer's process:

```
scala> val data = Array(8, 1, 2, 7, 9, 0, 5) // Step 1: set or load
data
data: Array[Int] = Array(8, 1, 2, 7, 9, 0, 5)
```

We have used a very small sample of seven numbers. This is a good enough size to understand the problem and appreciate the underlying issues with the proposed solution.

2. Sort the data in ascending order in the same process:

```
scala> val dataSorted = data.sorted // Step 2: sort the data
dataSorted: Array[Int] = Array(0, 1, 2, 5, 7, 8, 9)
```

Please note that `sort` is a very expensive operation, particularly when performed on a large dataset. We covered the performance characteristics of Scala's collection API in Chapter 1, *Scala Overview*. There are several choices of sort algorithms, because a balance is required between intermediate memory usage and overall compute time. For most sort algorithms, the worst case performance is *O(n log(n))*. This implies that the time it takes to sort the dataset is roughly proportional to the number of elements in the dataset.

3. Extract the `median` value as a midpoint from the sorted data:

```
scala> val median = dataSorted(dataSorted.size/2) // Step 3:
extract the median as midpoint from sorted data
median: Int = 5
```

This solution works well for small datasets, but it is unlikely to work successfully for large datasets for the following reasons:

- A single computer is unlikely to have enough RAM to hold all of this data in memory.
- The time it takes to compute is likely to be very high because of the limited number of processing cores in a single computer. Many sorting algorithms can be parallelized; however, each thread still needs a CPU core in order to execute.
- The sorting algorithm itself is going to require a significant amount of intermediate memory to perform its computations.

We are essentially facing two limitations of a single computer:

- Limits on the total amount of main memory
- Limits on the number of cores

It is certainly possible to solve such problems on a single supercomputer with huge amounts of RAM and an equally huge number of cores, but the cost of doing so would be too high and wouldn't be economically viable for any enterprise.

The volume of data significantly changes how we solve such problems and forces us to explore alternative algorithms toward finding feasible solutions. Let's look at solving this problem in a scalable and cost-effective way. For solving real-world problems, it is very important to appreciate the following factors associated with data:

- A good understanding of the domain and context in which the problem is to be solved goes a long way in building cost-effective solutions.
- Analysis, discovery, and observations of some key properties of the associated real data is a must. The data generally tells a story and a lot can be learned from that.

Let's assume that the data in our current context has the following properties:

- All the records are integral numbers
- The total count records are of the order of billions
- The range of integral value is 1 to 1,000

For the sake of generation, we will use the notion of a record that contains related data points associated with an observation. Our record has only one data point, which can have only numeric values.

The Range property is a very significant property, because of the real-world nature of data. With this property in mind, we can look at solving this in a completely different way. We defined a range of 1 to 1,000, which implies that there are a significant number of repeated values in the dataset.

Some such properties could come from the business rules of the domain, while others can be discovered using frequency analysis on data. To get an intuition for this, let's look at a simple Scala example in Scala REPL:

1. Create a dataset of 20 random integers in the Range[0, 5) class:

```scala
scala> val data = Range(0, 20).map(i =>
scala.util.Random.nextInt(5)) // Step 1: Create dataset of 20 random
                              integers in range [0, 5)
data: scala.collection.immutable.IndexedSeq[Int] = Vector(3, 2, 4,
2, 2, 3, 3, 1, 2, 4, 3, 3, 0, 0, 0, 1, 1, 4, 0, 3)
```

We used the Scala Range class to generate 20 numbers from 0 to 19 and map each one of these to a random number from 0 to 4 using the scala.util.Random object's nextInt method. We get a Vector object consisting of 20 random numbers from 0 to 4.

> Please note that each run of the preceding code would produce a different result because of the randomizing involved.

2. Get the distinct values from the dataset:

```scala
scala> data.distinct
scala> val dataDist = data.distinct // Step 2: Get distinct values
dataDist: scala.collection.immutable.IndexedSeq[Int] = Vector(3, 2,
                                                            4, 1, 0)
```

We use the `distinct` method of the `Vector` object that returned a new vector consisting of the five distinct values. We can indeed see that distinct values are 0 through 4, as expected. This kind of data analysis is sometimes known as **frequency analysis**.

3. Get `counts` for each distinct value:

```scala
scala> val counts = data.map(i => (i,
1)).groupBy(_._1).mapValues(_.size) // Step 3: Get counts for each
distinct value

counts: scala.collection.immutable.Map[Int,Int] = Map(0 -> 4, 1 ->
3, 2 -> 4, 3 -> 6, 4 -> 3)
```

To get the counts for each distinct value, we first mapped each value to a pair of (value, 1), performed a group by the value component of (value, 1), and finally counted 1 in each value. We can now see exactly the amount of occurrences of each number. Similar to the analysis in the previous step, this kind of analysis is also known as **frequency analysis**.

In the preceding example, we intentionally limited values from in the range of 0 to 4, and generated a dataset of 20 such random numbers. We observed that the dataset has 5 distinct numbers, as expected. We also saw that some values are repeated more often than others; for example, value 3 is repeated six times.

A real-world example of such a property could be an online retail system tracking the number of item purchases per order. An order would have at least one item and no more than 1,000 items, assuming that the retailer limits the maximum size of the order to 1,000 items.

The most important takeaway from these observations is that the dataset has a few distinct values and cannot exceed 1,000 in total. With this in mind, we apply a different algorithm to compute the median:

1. For each distinct value, count the number of occurrences. Represent this as a pair of (value, count).
2. Sort (value, count) in ascending order by sorting on the value part.
3. Compute the range index of each distinct value. The smallest value starts with index 0 and the index is incremented by the number of occurrences of this value to represent the start of the next index. We then move on to the next smallest number and repeat this process.
4. Determine the left index of the midpoint. This is needed to handle the situation when the numbers of the records are even. If this is the case, we have two midpoints and we need to take the average of the two to compute the median.

5. Determine the right index of the midpoint.
6. Filter out the relevant ranges based on the left and right midpoint indexes.
7. Compute the median by averaging the filtered values.

Let's explore this algorithm in Scala REPL using a concrete example:

1. Create a sample dataset of 20 random integers in `range(0, 5)`:

```scala
scala> val data = Range(0, 20).map(i =>
scala.util.Random.nextInt(5)) // Step 1: Create dataset of 20
random integers in range [0, 5)

data: scala.collection.immutable.IndexedSeq[Int] = Vector(3, 2, 4,
2, 2, 3, 3, 1, 2, 4, 3, 3, 0, 0, 0, 1, 1, 4, 0, 3)
```

This dataset is the same dataset that was used in the previous example.

2. Count the number of occurrences for each distinct value:

```scala
scala> val counts = data.map(i => (i,
1)).groupBy(_._1).mapValues(_.size) // Count number of occurences
for each distinct value

counts: scala.collection.immutable.Map[Int,Int] = Map(0 -> 4, 1 ->
3, 2 -> 4, 3 -> 6, 4 -> 3)
```

The counts a `Map` object and it has a mapping from the value to the number of its occurrences.

3. Sort the counts by the value part:

```scala
scala> val sortedCounts = counts.toArray.sortBy(_._1) // Sort by
value

sortedCounts: Array[(Int, Int)] = Array((0,4), (1,3), (2,4), (3,6),
(4,3))
```

We used the `sortBy` method array to perform the sort. The method accepts an argument that defines the field to be used for sorting. We used the value part by using the _._1 Scala notation. This is shorthand for indicating the first element of the incoming object. In this case, the incoming object is a Scala tuple of [Int, Int].

4. Compute the range index for each distinct value:

```
scala> val valuesWithIndex = {var currentTotal = 0;
sortedCounts.map(kv => {val from = currentTotal; currentTotal +=
kv._2; val to = currentTotal -1; (kv._1, from, to)})} // Compute
range index for each distinct value
valuesWithIndex: Array[(Int, Int, Int)] = Array((0,0,3), (1,4,6),
(2,7,10), (3,11,16), (4,17,19))
```

This assigns an index range to each distinct value. The smallest value starts with an index of zero. This concept is discussed in more detail later in this chapter.

5. Compute the left index of the mid-point:

```
scala> val leftMidPointIdx = if (data.size % 2 == 0) data.size/2-1
else data.size/2 // left index of mid-point
leftMidPointIdx: Int = 9
```

Finding the median depends upon whether the dataset has an even or odd number of records. If it has an odd number of records, then there is only one mid-point. For an even number of records, there are two mid-points that have to be averaged to compute the median.

For an odd number of records, the left and right mid-points are the same.

6. Compute the right index of the mid-point:

```
scala> val rightMidPointIdx = data.size/2 // right mid-point index
rightMidPointIdx: Int = 10
```

The right index does not depend upon whether a dataset has an even or an odd number of records.

7. Filter out the entries that are not in the range of the left and right mid-points:

```
scala> val midPoints = valuesWithIndex.filter(v => (leftMidPointIdx
>= v._2 && leftMidPointIdx <= v._3) || (rightMidPointIdx >= v._2 &&
rightMidPointIdx <= v._3)) // Filter out the relevant distinct
values
midPoints: Array[(Int, Int, Int)] = Array((2,7,10))
```

From the `valuesWithIndex` method, we have filtered out all of the entries that are not within the range of the left and right mid-points. This must provide us with either one entry or two entries. In this specific case, we got only one entry (2, 7, 10) because of the left mid-point index 9 and right mid-point index 10. Both are within the range of 7 to 10.

8. Compute the `median` value from the mid-points:

```scala
scala> val median: Double = midPoints.map(_._1).sum/midPoints.size
// Step 7: Compute median by taking average

median: Double = 2.0
```

We take the average of the mid-point values to compute the median value. This is required in order to account for the possibility of two entries in mid-points as a result of an even number of records and each mid-point falling in different ranges.

We can easily represent the preceding median derivation logic as a reusable Scala function by defining the following function:

- `getMedian`:
  - **Input**: An array consisting of a pair of value and associated count
  - **Output**: Median value

```scala
def getMedian(counts: Array[(Int, Int)]): Double = {
  val totalCount = counts.map(_._2).sum // Total number of records
  val sortedCounts = counts.sortBy(_._1) // Sort by value
  // Compute range index for each distinct value
  val valuesWithIndex = {
    var currentTotal = 0
    sortedCounts.map(kv => { val from = currentTotal; currentTotal
+= kv._2; val to = currentTotal - 1; (kv._1, from, to) })
  }
  // left index of mid-point
  val leftMidPointIdx = if (totalCount % 2 == 0) totalCount / 2 - 1
else totalCount / 2
  val rightMidPointIdx = totalCount / 2 // right mid-point index
  // Filter out the relevant distinct values
  val midPoints = valuesWithIndex.filter(v => (leftMidPointIdx >=
v._2 && leftMidPointIdx <= v._3) || (rightMidPointIdx >= v._2 &&
rightMidPointIdx <= v._3))
  midPoints.map(_._1).sum / midPoints.size // Compute median by
taking average
}
```

The preceding Scala function is generalized enough, and can be reused once we are able to count distinct values in our dataset. The function should work on a single machine as long as the requirements for data properties are met; that is, our dataset has few distinct values.

Let's look at what is going on here in more detail. We can envision our dataset laid out in a linear fashion as follows:

```
// Sorted numbers
0, 0,  0,  0,  1,  1,  1,  2,  2,  2,  2,  3,  3,  3,  3,  3,  3,  4,  4,
4
// Corresponding index
0, 1,  2,  3,  4,  5,  6,  7,  8,  9, 10, 11, 12, 13, 14, 15, 16, 17, 18,
19
```

We are reducing the aforementioned information into the following condensed form:

Value	Start Index	End Index
0	0	3
1	4	6
2	7	10
3	11	13
4	14	19

Once we have the aforementioned information, we can easily compute the median value. This algorithm is significantly different from the previously discussed simpler version. Now the question is, how able is this new algorithm to scale for large datasets? This is primarily because counting distinct values has two distinct properties that make this algorithm parallelizable:

- We can easily combine two or more such results into a single result if the individual results are computed from two mutually exclusive slices of the dataset.
- The dataset can be sliced in any way and any number of pieces, we would still get the same computed final result.

To verify this hypothesis, let's perform the following in Scala REPL:

1. Create a sample dataset of 20 random integers in `range` `[0, 5)`:

```
scala> val data = Range(0, 20).map(i =>
scala.util.Random.nextInt(5)) // Step 1: Create dataset of 20
random integers in range [0, 5)

data: scala.collection.immutable.IndexedSeq[Int] = Vector(3, 2, 4,
2, 2, 3, 3, 1, 2, 4, 3, 3, 0, 0, 0, 1, 1, 4, 0, 3)
```

This dataset is the same dataset that was used in the previous two examples.

2. Split our data into two parts, with even numbers in one part and odd numbers in the other part:

```
scala> val (evens, odds) = data.partition(_ % 2 == 0) // Split data
into even and odd

evens: scala.collection.immutable.IndexedSeq[Int] = Vector(2, 4, 2,
2, 2, 4, 0, 0, 0, 4, 0)

odds: scala.collection.immutable.IndexedSeq[Int] = Vector(3, 3, 3,
1, 3, 3, 1, 1, 3)
```

We use the `partition` method of `Vector` to split the data into two new vectors. The `evens` object contains all even numbers, while the `odds` object contains all odd numbers.

3. Compute the distinct counts for even numbers:

```
scala> val evenCount = evens.map(i => (i,
1)).groupBy(_._1).mapValues(_.size) // Compute distinct value count
for even

evenCount: scala.collection.immutable.Map[Int,Int] = Map(2 -> 4, 4
-> 3, 0 -> 4)
```

4. Compute the distinct count for odd numbers:

```
scala> val oddCount = odds.map(i => (i,
1)).groupBy(_._1).mapValues(_.size) // Step 3: Compute distinct
value count for odd

oddCount: scala.collection.immutable.Map[Int,Int] = Map(1 -> 3, 3
-> 6)
```

5. Combine the two results to create the final result:

```scala
scala> val combined = evenCount.union(oddCount).map(kv => (kv._1,
kv._2)).groupBy(_._1).mapValues(_.map(_._2).sum) // Step 4: Combine
the results of even and odd

combined: scala.collection.immutable.Map[Int,Int] = Map(0 -> 4, 1
-> 3, 2 -> 4, 3 -> 6, 4 -> 3)
```

We first took a union of `odd` and `even` counts and mapped each entry to a Scala tuple of `[Int, Int]` as a `(value, count)` pair. We then performed a `groupBy` method on the value part and mapped the values summing the counts.

The preceding code example proves that our hypothesis is correct. This can be confirmed by performing two or more random splits and by repeating similar steps. Once we have this condensed information that only consists of five entries, we can apply the previous algorithm, defined in the `getMedian` function, to get the median value. The main achievement of the aforementioned methodology is the ability to work on multiple smaller segments of data, compute intermediate results, and then combine these results to arrive at the same answer as when it was done traditionally.

Let's re-examine at these steps at a broader level to get a complete picture.

At a very high level, we are doing the following things:

1. We start with the data to be processed.
2. We then split the data into multiple segments or splits, S1 through Sn, where n is the number of segments. This assumes that data is splittable. An example of data that can be split easily is the CSV data format where a newline designates the start of a new record. A CSV format file of 10 million records can easily be split into 10 segments, with each segment having approximately 1 million records. This is possible because of the newline character being the CSV record delimiter.
3. Apply identical computation on each segment; the results of each individual segment are represented as R1 through Rn.
4. Combine the results for R1 through Rn to produce the final results.

The following diagram is a visual representation of the outlined steps and data flow:

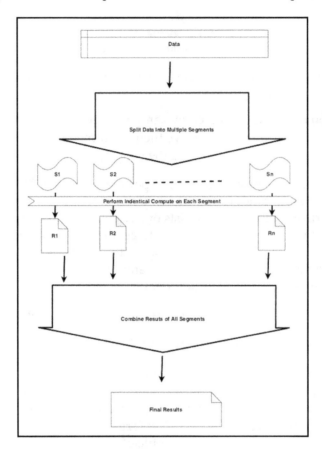

The previous steps are fundamental to processing data at scale. In order to solve problems that involve massive datasets, we need to think of data and computation as two separate entities. At the beginning of this book, we compared object-oriented programming and functional programming. Object-oriented programming treats data and computation as a single entity. On the other hand, functional programming treats these two as separate entities. Such algorithms are more in line with functional programming principles. That is the main reason why functional programming is ideally suited to solving such problems at a large scale.

Now, let's look at a complete example of solving this problem in Spark, which can leverage a large cluster of nodes to solve this problem at a massive scale. Spark uses the terminology **partition**, which is equivalent to a **split** or **segment** used in the earlier explanation. For demonstration purposes, we will use small datasets of 20 numbers and 2 partitions to begin with:

1. Start a new Spark shell at the command line:

```
$ spark-shell
```

2. Generate 20 random numbers in `range [0, 5)` with two partitions:

```
scala> // Generate 20 random numbers in range [0, 5) with 2
partitions

scala> val ds = spark.range(20). // 20 numbers
     | repartition(2). // 2 Partitions
     | map(i => scala.util.Random.nextInt(5)). // Randomize
     | cache() // Preserve random values
ds: org.apache.spark.sql.Dataset[Int] = [value: int]
```

We utilized Spark's `Range` function to generate `20 numbers`, instructed Spark to create two partitions, and then mapped each number to a random number between 0 to 4. This produces a Spark dataset of `Int`. Finally, we cached the dataset to preserve the values in the dataset during the session.

3. Map each partition to count distinct values:

```
scala> // Map each partition to count distinct values

scala> val dsWithCount = ds.mapPartitions(_.toArray.map(i => (i,
1)).groupBy(_._1).mapValues(_.size).toIterator)
dsWithCount: org.apache.spark.sql.Dataset[(Int, Int)] = [_1: int,
_2: int]

scala> // Step 3: Combine the results

scala> val combined = dsWithCount.rdd.reduceByKey(_+_).collect
combined: Array[(Int, Int)] = Array((4,3), (0,4), (2,2), (1,5),
(3,6))
```

4. Combine the results of both partitions:

```
scala> // Combine the results

scala> val combined = dsWithCount.rdd.reduceByKey(_+_).collect
combined: Array[(Int, Int)] = Array((4,3), (0,4), (2,2), (1,5), (3,6))
```

Please note that Spark is lazily evaluated, and this required us to use the cache API in step two. This is because of Spark's compute model, where the entire DAG is re-evaluated whenever an action is performed on the dataset. By caching this dataset, we are suggesting that Spark should evaluate the DAG for this dataset only once during the session. As we are generating random numbers, without the cache API usage, each Spark action on the dataset will produce a different result.

Let's check whether the aforementioned computation was performed correctly or not by examining the following data:

```
scala> ds.collect
res0: Array[Int] = Array(3, 1, 4, 0, 4, 1, 3, 2, 4, 0, 2, 3, 0, 1, 1, 3, 0,
3, 1, 3)
```

We can count manually to verify that the combined results are correctly computed.

Although this example was demonstrated using small numbers of data points and partitions, it can be made to work with massive datasets by having a large enough Spark cluster of computer nodes and tuning some of the parameters, such as the number of partitions. We can examine more details in Spark's UI by going to localhost:4040 and looking at the **Jobs** tab:

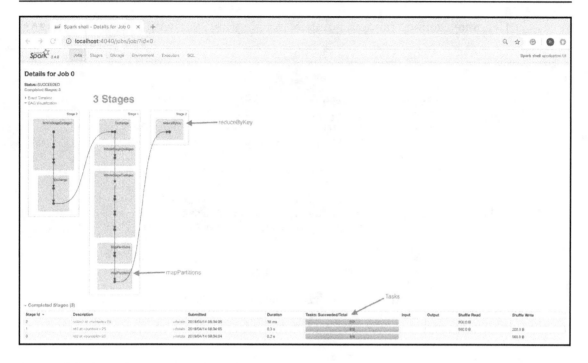

Some of the key observations from the UI are as follows:

- **Job 0** is the first **Job**.
- **Job 0** consists of three stages: **Stage 0**, **Stage 1**, and **Stage 2**.
- We can see the `mapParitions` and the `reduceByKey` transformations being performed on the dataset. This is triggered by the `collect` action.
- **Stage 0** consists of four tasks, **Stage 1** and **Stage 2** have two tasks each.

Now, let's rerun the same Spark recipe with the following variations:

- 1,000,000 records
- 10 partitions (100,000 records in each partition)

We will then observe Spark's behavior:

1. Start a new Spark shell at the command line:

    ```
    $ spark-shell
    ```

2. Generate 1 million random numbers in `range [0, 5)` with 10 Spark partitions:

```
scala> // Generate 1,000,000 random numbers in range [0, 5) with 10
partitions

scala> val ds = spark.range(1000000). // 1,000,000 numbers
     | repartition(10). // Partitions
     | map(i => scala.util.Random.nextInt(5)) // Randomize
ds: org.apache.spark.sql.Dataset[Int] = [value: int]

scala> // Step 2: Map each partition to count distinct values

scala> val dsWithCount = ds.mapPartitions(_.toArray.map(i => (i,
1)).groupBy(_._1).mapValues(_.size).toIterator)
dsWithCount: org.apache.spark.sql.Dataset[(Int, Int)] = [_1: int,
_2: int]

scala> // Step 3: Combine the results

scala> val combined = dsWithCount.rdd.reduceByKey(_+_).collect
combined: Array[(Int, Int)] = Array((0,199918), (1,200006),
(2,199896), (3,199636), (4,200544))
```

3. Map each partition to count distinct values:

```
scala> // Map each partition to count distinct values

scala> val dsWithCount = ds.mapPartitions(_.toArray.map(i => (i,
1)).groupBy(_._1).mapValues(_.size).toIterator)
dsWithCount: org.apache.spark.sql.Dataset[(Int, Int)] = [_1: int,
_2: int]
```

4. Combine the results from all partitions:

```
scala> // Combine the results

scala> val combined = dsWithCount.rdd.reduceByKey(_+_).collect
combined: Array[(Int, Int)] = Array((0,199918), (1,200006),
(2,199896), (3,199636), (4,200544))
```

Please note that we intentionally removed the caching of data to avoid memory issues.

Now, let's look in Spark UI again, at `localhost:4040`, and look at the **Jobs** tab:

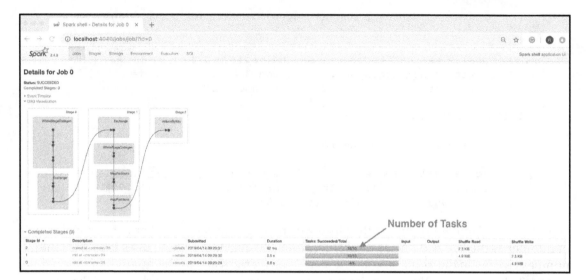

If we compare this UI result to the previous one, we can see a few changes:

- Even though the number of stages is still 3, the **stage 1** graph has changed slightly. This is because of the removal of the cache operation.
- **Stage 1** and **Stage 2** now have 10 tasks each. Since we have 10 partitions now, each task corresponds to a Spark partition and we now see 10 tasks instead of 2.

As a final review, let's look at a complete and simplified solution to this problem using Spark, by simplifying the map partition step significantly:

1. Start a new Spark shell at the command line:

   ```
   $ spark-shell
   ```

2. Generate 1 million random numbers in `range [0, 5)` with 10 partitions:

   ```
   scala> // Generate 1,000,000 random numbers in range [0, 5) with 10
   partitions

   scala> val ds = spark.range(1000000). // 1a: 1,000,000 numbers
        | repartition(10). // 1b: 10 Partitions
        | map(i => scala.util.Random.nextInt(5)) // 1c: Randomize
   ds: org.apache.spark.sql.Dataset[Int] = [value: int]
   ```

3. Map each `num` as a key-value pair of `(num, count)`, where `count = 1`:

```
scala> // Map each num as key value pair of (num, count) where
count = 1

scala> val dsWithCount = ds.rdd.map((_, 1))
dsWithCount: org.apache.spark.rdd.RDD[(Int, Int)] =
MapPartitionsRDD[8] at map at <console>:25
```

4. Combine the results:

```
scala> // Combine the results

scala> val combined = dsWithCount.reduceByKey(_+_).collect
combined: Array[(Int, Int)] = Array((0,199724), (1,199558),
(2,199909), (3,200306), (4,200503))
```

Step three is simplified significantly, since we are mapping each number to a key-value pair of number as the key and one as the value. Now, the question is, how does it scale as the number records in the `reduceByKey` step are exactly the same as the number of original records? The answer to this lies in how Spark's `reduceByKey` API works under the hood. Spark's `reduceByKey` API does the following:

- Apply `reduceByKey` for each partition locally and produce local results.
- Combine all local results from every partition to produce an aggregate result.

In the previous solution, we had used Spark's `MapPartitions` API to achieve this by computing results for each partition; however, we needed to bring the complete data from the partition in memory to compute that. We replaced them `MapPartitions` API with the much simpler `map` API of Spark. We can rely on Spark to perform all of the optimizations related to memory usage.

Let's see if we make this example work with a billion numbers in Spark. We will do essentially the same steps, except for a small modification in the usage of Spark's `range` API:

```
val ds = spark.range(0, 1000000000, 1, 1000). // 1,000,000,000 numbers
starting with 0 with 1000 partitions
```

We removed the repartitioning in the preceding step to avoid memory issues, because we are running Spark locally on a single machine. Instead, we asked Spark to create 1,000 partitions implicitly for us instead of using the `repartition` API explicitly. The `repartition` operation, if done explicitly on a large dataset, is a fairly expensive operation in terms of memory and network usage.

Let's run the example in the Spark shell and see whether it works:

1. Start a new Spark shell at the command line:

```
$ spark-shell
```

2. Generate 1 billion random number in the range of [0, 5) with 1000 partitions:

```
scala> // Generate 1,000,000,000 random numbers in range [0, 5)
1000 partitions

scala> val ds = spark.range(0, 1000000000, 1, 1000). //
1,000,000,000 numbers starting with 0 with 1000 partitions
    | map(i => scala.util.Random.nextInt(5)) // Randomize
ds: org.apache.spark.sql.Dataset[Int] = [value: int]
```

3. Map each num as a key-value pair of (num, count), where count = 1:

```
scala> // Map each num as key value pair of (num, count) where
count = 1

scala> val dsWithCount = ds.rdd.map((_, 1))
dsWithCount: org.apache.spark.rdd.RDD[(Int, Int)] =
MapPartitionsRDD[4] at map at <console>:25
```

4. Combine the results:

```
scala> // Combine the results

scala> val combined = dsWithCount.reduceByKey(_+_).collect
combined: Array[(Int, Int)] = Array((0,200012500), (1,199977942),
(2,200005787), (3,199988344), (4,200015427))
```

This actually worked, although it took a few minutes to complete. We can intuitively see that it correctly computed results, as we are generating random numbers from 0 to 4. Assuming a nearly uniform distribution of random numbers, each number should have occurred about 200 million times.

You can experiment with different combinations of the number of records, partitions, and range to get a good understanding of some of these principles. The following is the output with a range of (0, 20], using 1 billion numbers and 1,000 partitions:

```
scala> val ds = spark.range(0, 1000000000, 1, 1000). // 1a: 1,000,000,000
numbers starting with 0 with 1000 partitions
    | map(i => scala.util.Random.nextInt(20)) // 1b: Randomize
...
scala> val combined = dsWithCount.reduceByKey(_+_).collect
```

```
combined: Array[(Int, Int)] = Array((0,49990682), (1,49996451),
(2,50012140), (3,50004536), (4,49995865), (5,50006713), (6,50003562),
(7,50006684), (8,50006651), (9,50011185), (10,49993332), (11,49993691),
(12,50000625), (13,49989406), (14,49994246), (15,49994717), (16,50005893),
(17,49996304), (18,49997009), (19,50000308))
```

As can be seen in the preceding code, each number has roughly 50 million counts. From these experimentations, we can conclude that this solution is scalable, however, we need to realize that context is very important. In this case, our dataset is very large, however, the number of distinct values is relatively small. If the dataset had mostly unique values, this solution would not work and would run into the memory limitations of a single computer. With mostly unique values, the `reduceByKey` step would produce nearly the same number of records as the original dataset, and that is unlikely to fit into the memory on a single computer.

Working with data at scale has its challenges and requires a paradigm shift in working with such datasets. The traditional algorithms that work with small datasets need to be adapted and modified so that these can work on smaller slices of data, and are able to combine the results from smaller slices of in such a way that end result is the same.

Cost considerations

As the size of data grows, there are many factors to consider to manage costs effectively. Some of the costs associated with data are direct, while others are indirect. A clear and well-defined data strategy plays a central role in managing these costs and maximizing the value of data.

There are multiple points of view to consider when looking at the cost:

- Data storage
- Data governance

Data storage

Not at all data is created equal. Some types of data have more value than the others. The value of data might also be sensitive to its age and might start to decrease as the data ages. At the same time, some data is accessed more frequently than others. All of these factors, and many more, determine how the data is stored and will influence storage decisions.

Faster access storage would significantly increase the cost of storing data, but would also increase the speed of storing and retrieving processed data. An excellent example of this is **Solid State Storage (SSS)** technologies that provide high-performance data storage and retrieval, but cost significantly more than the more prevalent spindle disk-based magnetic storage. Magnetic disks are still the most cost-effective solutions for high-density data storage; however, due to the mechanical movements involved, it does add significant latency to data retrieval.

In the previous section, we looked at how working with data at scale requires us to adapt and change data processing algorithms. In fact, cost-effective processing requires data to be processed and stored in a certain way. In fact, when we have a network of computers processing a large dataset, the dataset must be accessible over the network. The dataset also has to be splittable in such a way that multiple workers could perform the computations on each assigned split or segment in a parallelized fashion. The computation logic is generally small in size and can be distributed over the network to multiple workers without incurring too much of a cost. On the other hand, if we have to move data around, the cost of doing so is significantly high because of the following reasons:

- Large datasets involve the movement of a huge amount of data over the network.
- The network is generally the slowest communication link when it comes to a computer's data processing.

To get a feel for how the network contributes to latency, please refer to `https://people.eecs.berkeley.edu/~rcs/research/interactive_latency.html`.

The key takeaways from the preceding link are as follows:

- It is always faster for a process to read the data from the main memory.
- SSDs provide the next fastest speed, followed by disks.
- Networks add additional latency and it progressively gets progressively worse as the physical separation becomes wider. For example, it is relatively faster when communication happens within the same data center, but becomes most latent on **wide area networks (WANs)**.

One other aspect of networking to realize is that it is a shared resource used for connecting multiple computers. Even though these networks might be very high-speed networks, their overall bandwidth is limited because of its shared nature.

When designing a cost-effective storage strategy for big data systems, networking topology must be kept in mind. Networking is the backbone of any distributed processing system, so it is impossible to avoid network usage completely; however, the overall system has been designed to reduce an excessive flow of data over the network, particularly during the data processing phase.

Apache Hadoop's distributed file system, **Hadoop Distributed File System** (**HDFS**), was designed with some of these concerns in mind. When storing a file, HDFS slices it multiple large chunks, typically of 128 MB in size. Each slice is stored on multiple nodes, and most typically on three nodes. The slices are somewhat randomly distributed on different nodes in the cluster. Furthermore, most Apache Hadoop clusters have all of their nodes located within a single data center, where these are connected by a high-speed network. It also supports the notion of rack awareness, where nodes located within the same rack can communicate with each other much faster than going across the rack.

When this data needs to be processed, Apache Hadoop MapReduce or Apache Spark plan the work in such a way that the worker and data are colocated as much as is possible. In essence, computation is moved to a worker node in the cluster where the data is also located.

Let's look at how this works in practice using a simple example:

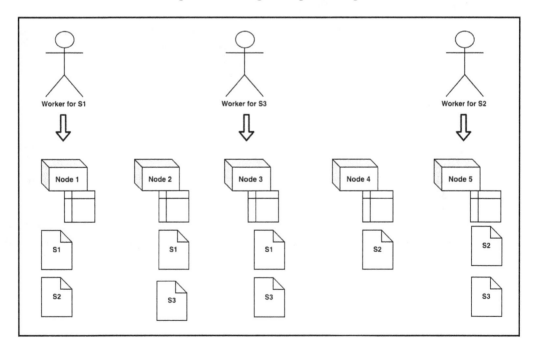

We have the following configuration:

- A five node cluster, with **Node 1** through **Node 5**, within a single data center connected through a high-speed network.
- Each node has local storage attached to it.
- We also have a file that has three slices, **S1** through **S3**.

Here is the distribution of the file's three slices on different nodes in the cluster:

Node 1	Node 2	Node 3	Node 4	Node 5
S1	S1	S1	S2	S2
S2	S3	S3		S3

The selected nodes in this example for processing are this dataset:

S1	S2	S3
Node 1	Node 5	Node 3

The arrangement of data slices to the assigned worker node avoids the usage of moving the source dataset around during processing and can be processed efficiently. Since we have three copies of each slice of data, this design also works well when there are node failures. For example, let's say **Node 1** dies because of some hardware failure-slice **S1** needs to be processed somewhere else. **Node 2** and **Node 3** have copies of slice **S1**. It can then easily use the following configuration to process this dataset:

S1	S2	S3
Node 2 (instead of Node 1)	Node 5	Node 3

Solutions such as Hadoop are great for on-premises data centers.

Now the question is, what happens when these data processing workloads run in the cloud, such as on **Amazon Web Services (AWS)** or **Google Cloud Platform (GCP)**? One of the greatest benefits of a cloud-based solution is on-demand computing, and having the flexibility to pay only for the computations and storage being used. The cloud is also a paradigm shift in itself and requires thoughtful architectural considerations in terms of how and where the data should be stored.

Earlier, we touched upon the point that some data is accessed more frequently than others. The cloud offers many great choices in this regard and can help manage not just stored data, but will also compute more efficiently, if designed properly.

AWS, for example, offers many choices in terms of data storage. For example, AWS's **Simple Storage Service (S3)** could be used to store data that can be accessed efficiently from within the AWS ecosystem. AWS S3 is an object store and is, by default, distributed over the network. AWS handles the network optimizations when S3 data is accessed from within the AWS ecosystem by big data frameworks such as Spark and MapReduce. For long-term storage, AWS Glacier provides a cost-effective solution; however, this is not suitable for frequent and low-latency access.

The data that needs to be accessed with low latency of an order of a few milliseconds requires very approach to storage. A key-value store, such as Apache HBase and Apache Cassandra, are some of the choices that can be used on-premises. For AWS, Dynamo DB is a great choice because it is fully managed and provides elasticity of usage and storage in a cost-effective way.

Data governance

Data governance is a very broad topic in itself. With the advent of big data, this has become one of the biggest challenges to deal with. There are strict government rules with respect to what type of data an organization can collect, store, and use. Many of these rules are country-specific and must be followed rigorously in order to stay in compliance and avoid severe penalties.

The problem that big data poses is that there is so much of it and it generally manifests all of its three V properties, namely, volume, variety, and velocity. Sensitive data must be secured at all phases and must have well-established access control mechanisms throughout the enterprise. In order for modern enterprises to innovate and succeed in today's business environment, employees need access to data. Some of this data could be sensitive. Having excessive and unnecessary access control to data might prevent an organization from innovating rapidly. On the other hand, having poor access control around data might lead to compliance violations.

The solution to this problem once again lies in having a well-defined data strategy, where data governance must be a first-class citizen. Whenever a new dataset is collected and stored, the following questions must be asked:

- Is the data relevant to the company's business? Collecting and storing data irrelevant to the business creates an unnecessary cost that can be avoided.
- Does the data contain sensitive information? If yes, then what is the mechanism to be used to secure this data in transit and at rest?

- How would the lineage of this data be maintained? For example, if some part of the source data is bad and gets corrected later on, how would the downstream system handle these corrections?
- What is the retention policy around this data? If the data needs to be purged, what must be done about the linked data stored in downstream systems?

These are only some of the questions that must be answered beforehand, and clear guidelines need to be defined to address these at the outset.

For large data volumes, one of the biggest problems is cleaning up data. Let's look at a specific example of an online retailer that tracks customer's activity on its website. In some countries, there is a law that authorizes a customer to have their data deleted once the customer closes his or her account. This cleanup action must be performed online within a designated period. This can be a very difficult problem to solve if the data storage and processing systems are not designed with this specific requirement in mind. A customer's data could be sitting in raw files, database tables, Excel files, and many more places. This is where strong data governance policy and lineage tracking comes into the picture. The system must be designed to facilitate the cleanup of a customer's data efficiently and correctly, without impacting the system's overall performance.

One common technique used to handle scenarios like this is to segregate the data that needs to be cleaned up, from that data that can remain immutable until it has reached its end of life cycle. Once again, to design a cost-effective data governance solution, knowledge of the business domain is essential, along with a good understanding of the underlying tools needed to support it.

Reliability considerations

Processing large datasets requires reliability to be looked at from a slightly different point of view. It is quite common to have a small percentage of errors in such large datasets. An acceptable error tolerance level can only be defined by business rules. Large datasets are generally processed by a network of computers, where failures are more common compared to processing on a single computer. In this section, we will look at the following aspects of error handling:

- Input data errors
- Processing failures

Input data errors

As a general guideline, it is crucial to measure and monitor the number of errors in the input data over time. If the quality of the input data is bad, then any analysis performed and conclusions drawn on that data would be unreliable. In fact, if the downstream systems consuming that data are unaware of the bad quality of the data, this might have undesirable results, creating a chain reaction of failures.

The data coming from external sources, that is, sources outside of the enterprise, in particular, must have very well defined rules with respect to validation and acceptance of data. Just imagine an FTP data file that was sent by a third party that got corrupted in transit. Processing this corrupted data can have unintended consequences. One way to protect against such data corruption is to have an associated checksum file and ensure that the data file has the expected checksum.

It is also possible that some of the records in the input dataset might have missing values or incorrect data types. It might be acceptable to have a few hundred bad records in a dataset consisting of billions of records. Again, this all depends on the business rules and the context that this data applies to.

Let's look at an example in Spark on how to handle and measure such data errors:

1. Start a new Spark shell at the command line:

   ```
   $ spark-shell
   ```

2. Import Spark's `functions` package:

   ```
   scala> import org.apache.spark.sql.functions._
   import org.apache.spark.sql.functions._
   ```

3. Import Spark's `types` package:

   ```
   scala> import org.apache.spark.sql.types._
   import org.apache.spark.sql.types._
   ```

4. Create a dummy dataset with good and bad records:

   ```
   scala> val df = List("1", "one", "2", "3", "4").toDF // Create a
   dummy dataset of records with one column called value. Some records
   have a valid number while a few have an invalid string.
   df: org.apache.spark.sql.DataFrame = [value: string]
   ```

We will get a Spark DataFrame here that has only one field. The field name is `value` and the datatype is `string`.

5. Add a new `int_value` column to the dataset by casting the original value to the integer type:

```
scala> val dfWithInt = df.withColumn("int_value",
col("value").cast(IntegerType)) // Create a new dataset by adding
int_value column that converts value column to int.
dfWithInt: org.apache.spark.sql.DataFrame = [value: string,
int_value: int]
```

The `cast` function is defined in Spark's functions package, while `IntegerType` is defined in Spark's `types` package.

6. Add a new Boolean column called `has_error` to indicate whether there was an error in casting from `string` to `integer`:

```
scala> val dfWithError = dfWithInt.selectExpr("value", "int_value",
"(value is not null and int_value is null) has_error") // Add a new
column called has_error which set to true if invalid data type.
dfWithError: org.apache.spark.sql.DataFrame = [value: string,
int_value: int ... 1 more field]
```

Spark DataFrame's `selectExpr` method lets us use a SQL-like expression as the parameters. This produces a new DataFrame by applying each of the supplied expressions.

7. Display the contents of the new dataset:

```
scala> dfWithError.show // Show the dataset
+-----+---------+---------+
|value|int_value|has_error|
+-----+---------+---------+
|    1|        1|    false|
|  one|     null|     true|
|    2|        2|    false|
|    3|        3|    false|
|    4|        4|    false|
+-----+---------+---------+
```

In the preceding example, we created a new indicative column that can represent whether an error occurred during conversion from string to integer. Please note that Spark silently set the value of the row in error to a null value. We can easily count the number of records in error:

```
scala> dfWithError.where("has_error").count
res4: Long = 1
```

Although the preceding example is fairly simple, it demonstrates how to measure errors using a few lines of code. This principle can easily be applied while processing large datasets where errors are tracked. In fact, in many cases, it is worthwhile storing such information along with processed data, so that the user of this data can make a decision on whether to use this dataset or discard a subset of records with bad data.

Processing failures

As highlighted earlier in this chapter, processing failures are quite common in distributed systems processing large datasets. The time it takes to process large datasets could be significant, and, if the processing is interrupted due to failures, it would extend the time it takes to complete processing.

Processing frameworks such as Apache Hadoop MapReduce and Apache Spark handle many of the system failures automatically, by retrying the failed portion of work on a different node. This is a very powerful feature of these frameworks that greatly simplifies a programmer's task. This benefit, however, comes with a few responsibilities:

- The work or computation to be performed has to be idempotent. What this really means is that, given an input to the worker, it must always produce the same output.
- The computation must not depend on the state of an external system.

A concrete example of this is a computation that updates a database table by incrementing a counter as part of its work. Let's say that, immediately after it has updated the database table, the node on which it is running fails. MapReduce or Spark will try this work on some other node on the cluster and healthy. This will cause the database table to be updated again and the counter will be incremented one more time. This type of processing is not idempotent and must not be performed on systems such as MapReduce and Spark.

Summary

In this chapter, we looked at working with data at scale. Working with large datasets requires a paradigm shift in how the data is processed. Traditional methods that work with smaller datasets generally don't work well with large datasets, because these are designed to work on a single computer. These methods need to be re-engineered to work effectively with large datasets. For scalability, we need to turn to distributed computing; however, this introduces significant additional complexity because of the network being involved, where failures are more common. Using good, time-tested frameworks, such as Apache Spark, is the key to addressing these concerns.

Another Book You May Enjoy

If you enjoyed this book, you may be interested in these other books by Packt:

Scala Machine Learning Projects
Md. Rezaul Karim

ISBN: 9781788479042

- Apply advanced regression techniques to boost the performance of predictive models
- Use different classification algorithms for business analytics
- Generate trading strategies for Bitcoin and stock trading using ensemble techniques
- Train **Deep Neural Networks** (**DNN**) using H2O and Spark ML
- Utilize NLP to build scalable machine learning models
- Learn how to apply reinforcement learning algorithms such as Q-learning for developing ML application
- Learn how to use autoencoders to develop a fraud detection application
- Implement LSTM and CNN models using DeepLearning4j and MXNet

Leave a review - let other readers know what you think

Please share your thoughts on this book with others by leaving a review on the site that you bought it from. If you purchased the book from Amazon, please leave us an honest review on this book's Amazon page. This is vital so that other potential readers can see and use your unbiased opinion to make purchasing decisions, we can understand what our customers think about our products, and our authors can see your feedback on the title that they have worked with Packt to create. It will only take a few minutes of your time, but is valuable to other potential customers, our authors, and Packt. Thank you!

Index

www.ingramcontent.com/pod-product-compliance
Lightning Source LLC
Chambersburg PA
CBHW080628060326
40690CB00021B/4855